Ordovician K-bentonites of Eastern North America

Dennis R. Kolata
Illinois State Geological Survey
615 East Peabody Drive
Champaign, Illinois 61820

Warren D. Huff
Department of Geology
547 Geology/Physics Building
University of Cincinnati
Cincinnati, Ohio 45221

and

Stig M. Bergström
Department of Geology and Mineralogy
155 South Oval Mall
Ohio State University
Columbus, Ohio 43210

SPECIAL PAPER
313
1996

Copyright © 1996, The Geological Society of America, Inc. (GSA). All rights reserved. GSA grants permission to individual scientists to make unlimited photocopies of one or more items from this volume for noncommercial purposes advancing science or education, including classroom use. Permission is granted to individuals to make photocopies of any item in this volume for other noncommercial, nonprofit purposes provided that the appropriate fee ($0.25 per page) is paid directly to the Copyright Clearance Center, 27 Congress Street, Salem, Massachusetts 01970, phone (508) 744-3350 (include title and ISBN when paying). Written permission is required from GSA for all other forms of capture or reproduction of any item in the volume including, but not limited to, all types of electronic or digital scanning or other digital or manual transformation of articles or any portion thereof, such as abstracts, into computer-readable and/or transmittable form for personal or corporate use, either noncommercial or commercial, for-profit or otherwise. Send permission requests to GSA Copyrights.

Copyright is not claimed on any material prepared wholly by government employees within the scope of their employment.

Published by The Geological Society of America, Inc.
3300 Penrose Place, P.O. Box 9140, Boulder, Colorado 80301

Printed in U.S.A.

GSA Books Science Editor Abhijit Basu

Library of Congress Cataloging-in-Publication Data
Kolata, Dennis R.
    Ordovician K-bentonites of eastern North America / Dennis R.
Kolata, Warren D. Huff, and Stig M. Bergström.
      p.   cm. -- (Special paper ; 313)
    Includes bibliographical references.
    ISBN 0-8137-2313-2
    1. Bentonite--North America.   2. Geology, Stratigraphic-
-Ordovician.   3. Geology--North America.   I. Huff, Warren D., 1937-
 . II. Bergström, Stig M.   III. Title.   IV. Series: Special papers
(Geological Society of America) ; 313.
QE391.B55K65   1996
553.6'1--dc20   96-34502
                                                                        CIP

Cover: The Deicke K-bentonite Bed in the Eggleston Formation near Hagan, Lee County, Virginia. Photograph taken by Stig M. Bergström in 1975.

10 9 8 7 6 5 4 3 2 1

Contents

Abstract 1

Introduction and Purpose 1

Geologic Setting 2

Field Identification of K-bentonites 3

Previous Investigations 3
    1920s and Early 1930s 3
    Mid-1930s Through Late 1970s 5
    1980s to Present 6

Mineralogical and Chemical Characteristics 7

Clay Mineralogy 7
    Age of Illitization 8
    Layer Charge of I/S in K-bentonites 9
    High Resolution Transmission Electron Microscopy 11
    Petrography 12
    Chemistry 12
    Chemical Fingerprinting 13

Biostratigraphic Framework 13

K-bentonite Distribution in Eastern North America 14
    Upper Mississippi Valley 15
    Eastern Missouri 20
    Illinois Basin 22
    Central and Northern Kentucky and Southern Ohio 25
    Central Tennessee 28
    Southernmost Appalachians 30

Southwestern Virginia and Eastern Tennessee 32
    Western Outcrop Belt 32
    Central Outcrop Belt 33
    Eastern Outcrop Belt 34
Northern Virginia and Eastern West Virginia 35
Central Pennsylvania 37
New York 38
Southern Ontario 41
Michigan Basin 44
St. Lawrence Lowlands and Quebec 46
Western Newfoundland 46
Eastern Oklahoma–Arkoma Basin 48
Western Iowa, Eastern Kansas, and Nebraska 48
Texas 49

Regional Correlation of K-bentonites 50
    Ibexian Series 51
    Whiterockian Series 51
    Mohawkian Series 51
        Deicke K-bentonite Bed 52
        Millbrig K-bentonite Bed 53
        Other Widespread Mohawkian K-bentonite Beds 54
    Cincinnatian Series 56

Regional Cross Sections 56
    Eastern Missouri–Southeastern Minnesota Cross Section 56
    Eastern Missouri–Southern Ontario Cross Section 57
    Eastern Missouri–Eastern Ohio Cross Section 58
    Eastern Missouri–Western Virginia Cross Section 58
    Southern Ohio–Western Virginia Cross Section 59
    Northeastern Kentucky–Northeastern Alabama Cross Section 60

Tectonic Setting and Origin of K-bentonites 60
    Composition of Parental Magmas 61
    Tectonic Setting 63

Event Stratigraphic Implications 64

Summary and Conclusions 66

Acknowledgments 68

Appendix 1. Locality Register 69

References Cited 76

Contents

Plates
(in pockets)

Plate 1. Correlation of Ordovician K-bentonites in eastern North America

Plate 2. Wireline-log cross section from eastern Missouri to southern Minnesota showing correlation of key Mohawkian K-bentonite beds

Plate 3. Wireline-log cross section from eastern Missouri to southwestern Ontario showing correlation of key Mohawkian K-bentonite beds

Plate 4. Wireline-log cross section from eastern Missouri to eastern Ohio showing correlation of key Mohawkian K-bentonite beds

Plate 5. Wireline-log cross section from eastern Missouri to western Virginia showing correlation of key Mohawkian K-bentonite beds

Plate 6. Wireline-log cross section from southern Ohio to western Virginia showing correlation of the Hagan K-bentonite Complex

Plate 7. Wireline-log cross section from northern Kentucky to northeastern Alabama showing correlation of key Mohawkian K-bentonite beds

Plate 8. Diagrammatic cross section extending from outcrops at St. Paul, Minnesota, to western Virginia

Ordovician K-bentonites of eastern North America

ABSTRACT

The Ordovician stratigraphic succession of eastern North America contains at least 60 altered volcanic ash beds, K-bentonites, one or more of which are distributed over an area of 1.5 million km^2. The beds range in age from Ibexian to Cincinnatian with the greatest concentration in mid-Mohawkian strata. Most K-bentonites are not widely distributed, but a few can be correlated for hundreds, or even thousands, of kilometers by chemical fingerprinting techniques, tracing on wireline logs, and matching of detailed outcrop descriptions.

The thickest and most widespread beds include the mid-Mohawkian Hockett (new), Ocoonita (new), Deicke, Millbrig, and Dickeyville K-bentonites, in ascending order. These beds, comprising the Hagan K-bentonite Complex, can be traced confidently in outcrop and on wireline logs from the Sevier Basin of western Virginia and eastern Tennessee westward and northward into the Illinois and Michigan Basins. The Deicke, Millbrig, and Dickeyville extend even farther northwestward into the Mississippi Valley, a total distance of approximately 1,300 km. Evidence suggests that the Deicke and Millbrig may extend into the Arkoma Basin of eastern Oklahoma.

Overall, the Ordovician K-bentonites are thicker, coarser, and more numerous toward the central and southern Appalachians, suggesting that the source volcanoes were situated in the region between what are now Alabama and Pennsylvania. Major- and trace-element analyses of whole rock K-bentonite samples indicate that the parental magmas consisted of a calc-alkaline suite ranging through andesite, rhyodacite, trachyandesite and rhyolite. Furthermore, the chemical compositions indicate a tectonomagmatic setting characterized by destructive plate-margin volcanics. Large volumes of volcanic ash ejected from plinian and co-ignimbrite eruptions along the active margin were carried by the prevailing southeast tradewinds for hundreds of kilometers northwestward and were deposited in shallow cratonic seas.

... [T]his ancient volcano exploded in one of the greatest eruptions which, so far as records are decipherable, we know the earth has undergone.

Wilbur A. Nelson (1922)

INTRODUCTION AND PURPOSE

During Ordovician time, the southeastern margin of Laurentia was characterized by explosive volcanic activity apparently involving subduction of the Iapetus Ocean crust and collision of microcontinents. Volcanic ash was discharged episodically into the atmosphere, and some was carried by wind currents for hundreds of kilometers into the interior of Laurentia where it fell into widespread cratonic seas. The Ordovician stratigraphic succession in eastern North America contains at least 60 volcanic ash beds, now altered to potassium-rich clay beds called K-bentonites. They also are referred to in the literature as bentonites or metabentonites.

The purpose of this book is to (1) summarize the mineralogies and chemical compositions that help distinguish individual beds and provide information regarding the tectonomagmatic setting of the source volcanoes, (2) document the geographic and stratigraphic distribution of the 60 or more Ordovician K-bentonites in eastern North America, (3) determine the relative

Kolata, D. R., Huff, W. D., and Bergström, S. M., 1996, Ordovician K-bentonites of eastern North America: Boulder, Colorado, Geological Society of America Special Paper 313.

positions of K-bentonites within an established biostratigraphic framework, and (4) determine which beds or bed complexes have potential event-stratigraphic significance. Results presented here represent more than 20 years of research, including sampling in the major outcrop belts and tracing individual K-bentonites through the subsurface to determine the lateral continuity of beds. We have drawn on a wide variety of information in published journals as well as industry reports, theses, state and provincial geologic records, and open-file reports.

Ordovician K-bentonites are a potential source of diverse geologic information. First, because they were deposited in a geologic instant over large areas, they constitute nearly isochronous rock units useful in precise correlations applicable to biogeographic, paleogeographic, paleoecologic, and sedimentologic investigations on both local and regional scales. The K-bentonites provide a precise temporal framework useful in determining coeval lithofacies, relative age and extent of unconformities, relative rates of sedimentation, timing of basin subsidence and uplift of arches and domes. Second, some of the volcanogenic crystals (zircon, apatite, biotite, feldspar) contained in K-bentonites are useful for isotopic dating. Third, the beds represent the distal, glass-rich portions of fallout ash derived from collision-zone explosive volcanism, and hence they have tectonomagmatic significance. Fourth, it is likely that such thick and widespread ash deposits and the presence of large amounts of dust in the atmosphere would have had a profound effect on the biota and may have produced climatic perturbations. A first step in understanding these aspects of Ordovician geologic history is to develop a regional K-bentonite stratigraphy, which is one of the purposes of this book.

Although it has been known since the 1920s that K-bentonites are widespread in Ordovician rocks of eastern North America, most attempts to correlate individual beds have been restricted to local outcrop belts. Some of the major problems often cited in regional correlations include (1) the occurrence of numerous beds in some regions (e.g., 30 beds in the Middle Ordovician Flat Creek Shale Member of the Utica Shale in the Mohawk Valley of New York; Charles E. Mitchell, 1994, personal communication), (2) the apparent mineralogic similarity of K-bentonites, and (3) the suspected lack of continuity of individual beds over long distances. Recent advances in the chemical characterization (Huff, 1983; Kolata et al., 1986, 1987; Delano et al., 1994) and study of phenocryst mineralogy (Haynes, 1992, 1994) provide a relatively high degree of accuracy in distinguishing individual beds. Furthermore, hundreds of wireline logs recording the presence of K-bentonites in the subsurface are available from widely distributed drill holes that penetrate the Ordovician succession (Huff and Kolata, 1990). Consequently, it is now possible to correlate individual beds confidently over long distances using a combination of chemical fingerprinting, tracing on wireline logs, and detailed outcrop descriptions.

This is a progress report that we hope will provide a framework for future investigations of North American Ordovician K-bentonites. Much additional work is needed, however, before the significance of these unusual rocks is fully understood.

GEOLOGIC SETTING

This book focuses on the Ordovician rocks in the eastern half of North America extending from the Appalachians of the United States and maritime Canada westward to the central part of the continent. Ordovician strata are widely distributed and range in thickness from approximately 300 m in the Upper Mississippi Valley region to about 3,000 m in the central Appalachian Mountains (Cook and Bally, 1975). Throughout the Ordovician, K-bentonite beds are present, the greatest concentration occurring in Mohawkian strata (Bergström, 1989; Huff et al., 1992). The beds are present in outcrop and/or the subsurface from near Le Mars in northwestern Iowa to Ardmore, Oklahoma, to Birmingham, Alabama, to Quebec City, Canada (Fig. 1). Small outliers containing K-bentonite beds are also known from the region of Port-au-Port in western Newfoundland. Ordovician K-bentonites range in thickness from a few centimeters in the Mississippi Valley to more than a meter for some beds in the southern Appalachians. Generally, the beds are thicker, coarser, and more numerous from west to east. The area covered by one or more beds is calculated at a minimum of approximately 1.5 million km^2. This does not include parts of Texas, Oklahoma, Nebraska, and Kansas where Ordovician K-bentonites are suspected to occur in the subsurface but have not been confirmed. Ordovician rocks are particularly well exposed around the margins of the Canadian Shield from the Upper Mississippi Valley across the Great Lakes into western Newfoundland and along the crests of the major arches and domes as well as in the fold and thrust belt of the Appalachian Mountains (Fig. 2). The western boundary of the study area is marked by the erosional featheredge of Ordovician strata at the eastern margin of the Transcontinental Arch, a linear positive feature that extended from central Minnesota into New Mexico during the Ordovician.

Lithologic, paleobiogeographic, and paleomagnetic data suggest that the proto–North American continent, Laurentia (Fig. 3), straddled the equator throughout most of the Ordovician (Ross, 1976; McKerrow et al., 1991). Furthermore, the southeastern margin of Laurentia underwent a long and complex history of rifting and subduction during the Ordovician. According to the widely cited Paleozoic plate reconstructions by McKerrow et al. (1991), small slivers of continental crust became detached from the southeastern margin of Laurentia during the Tremadoc, resulting in the opening of the Neo-Iapetus Ocean. The southeastern boundary of the Neo-Iapetus converted almost immediately into a northwest facing arc and subduction zone. The arc then collided diachronously with Laurentia causing successive orogenies from Tremadoc through Caradoc time. Volcanic ash from explosive eruptions associated with the arc/continent collisions was carried by prevailing winds northwestward, fell into shallow cratonic seas (Fig. 3) and subsequently altered to K-bentonite.

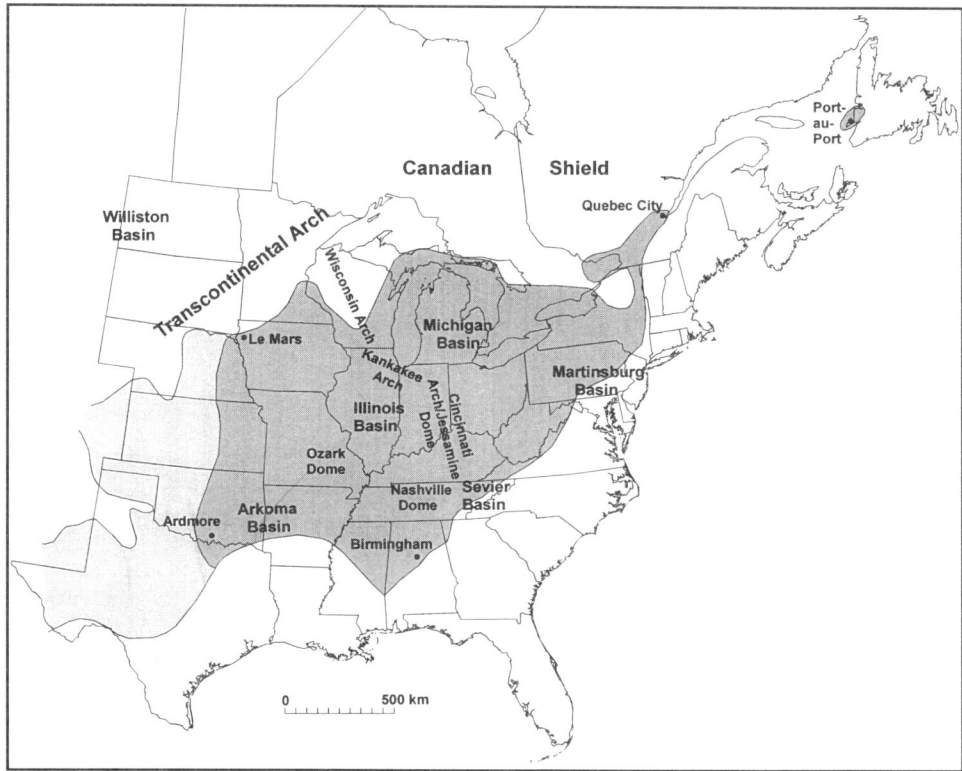

Figure 1. Areal distribution of Ordovician K-bentonite beds (shaded) and major structural features in eastern North America. K-bentonites are suspected to be present also in the subsurface from Nebraska to Texas (lightly shaded).

FIELD IDENTIFICATION OF K-BENTONITES

Most K-bentonites are characterized by a soapy to waxy texture when wet and a wide range of colors including light to dark shades of gray, buff, orange, and green depending upon the composition, activity of circulating ground water and local weathering processes. In cores and underground exposures where there has been less oxidation of iron, K-bentonites are more often green to greenish gray to bluish gray in color. Some K-bentonites contain conspicuous and abundant euhedral to anhedral biotite flakes. The typical appearance of a K-bentonite bed in outcrop is that of a fine-grained clay-rich band that has extruded by static load from the enclosing clastic or carbonate rocks. The bed then weathers rapidly and is thus recessed into the rock face. In limestone successions, a layer of chert commonly underlies K-bentonite beds that are greater than 50 cm thick. When K-bentonite beds are exposed in weathered outcrops, they commonly are overgrown with trees and shrubs growing in the moist, nutrient-rich clay. Some clay-rich shales closely resemble K-bentonites in the field but can be distinguished in the laboratory by X-ray diffraction analyses of oriented samples. Evidence of a volcanic origin of K-bentonites includes the areal persistence of individual beds, presence of locally preserved relict glass shards, and abundant euhedral crystals of zircon, feldspar, apatite, and biotite. Locally, the volcanic ash has been altered to fine-grained authigenic K-feldspar consisting of indurated beds that commonly are pinkish gray in color.

PREVIOUS INVESTIGATIONS

The study of Ordovician K-bentonites can be divided into three primary periods of research activity: (1) 1920s and early 1930s—recognition that K-bentonite beds are altered volcanic ash having event-stratigraphic significance, (2) mid-1930s to late 1970s—discovery of K-bentonites at numerous localities in eastern North America and a few attempts to correlate regionally, and (3) 1980s to the present—use of a variety of geochemical and mineralogical techniques to determine the genesis, distribution, and ages of K-bentonites.

1920s and early 1930s

The first report of an Ordovician K-bentonite in eastern North America was made by Ulrich (1888) who described a thick bed of clay in the upper part of what is now known as the Tyrone Limestone, near High Bridge, Kentucky. The volcanic origin and significance as a widespread marker bed, however, was not recognized until Nelson (1921, 1922) chemically analyzed a K-bentonite from central Tennessee and showed its sim-

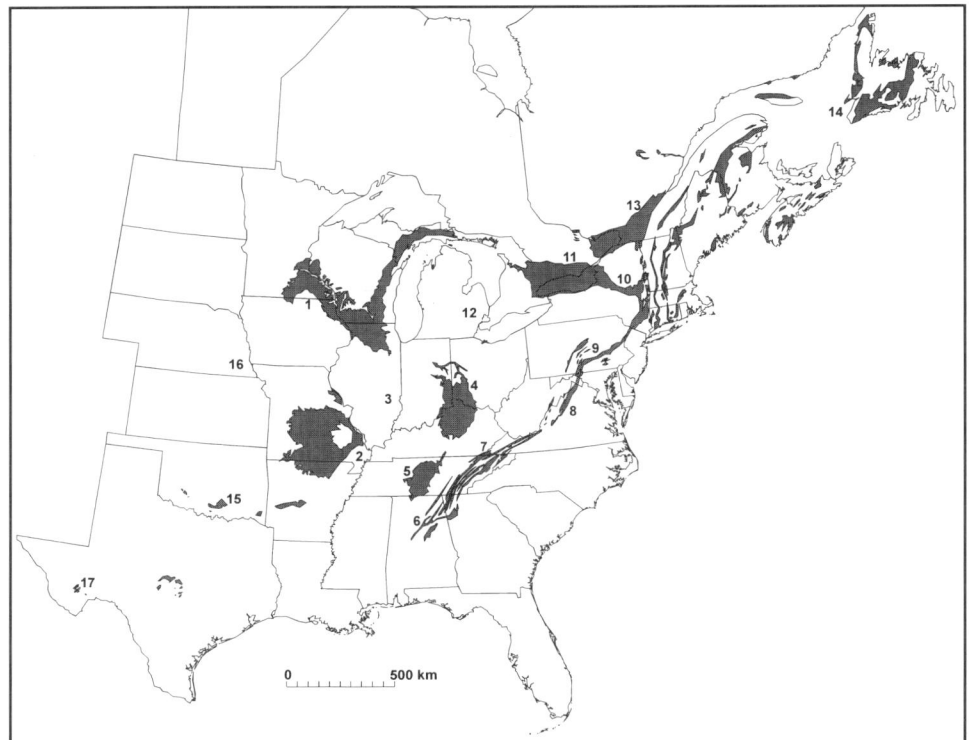

Figure 2. Areal distribution of Ordovician outcrops in eastern North America (shaded). Numbers correspond to outcrop belts, significant subsurface regions and cratonic basins emphasized in the text: 1—Upper Mississippi Valley, 2—eastern Missouri, 3—Illinois Basin, 4—north-central Kentucky, eastern Indiana and southern Ohio, 5—central Tennessee, 6—southernmost Appalachians, 7—southwestern Virginia and eastern Tennessee, 8—northern Virginia and eastern West Virginia, 9—central Pennsylvania, 10—upstate New York, 11—southern Ontario, 12—Michigan Basin, 13—St. Lawrence Lowlands and southern Quebec, 14—western Newfoundland, 15—Arbuckle Mountains and Arkoma Basin of eastern Oklahoma, 16—western Iowa and eastern Kansas and Nebraska, and 17—Marathon region of southwest Texas.

ilarity to Cretaceous and Tertiary bentonites of the western interior United States. Nelson (1922) correlated the bed from Singleton, Tennessee, south to Birmingham, Alabama, and north to High Bridge, Kentucky, covering a total distance of 540 km. He suggested that the ash may have covered a wide area of eastern North America from the Appalachians to Missouri and Arkansas. Nelson (1921) further speculated that the source volcano was situated on an island in northeastern Kentucky and that approximately 66 cubic miles (275 km^3) of ash were deposited in the surrounding sea.

Reports of K-bentonites in other regions of the United States began to appear in the literature soon after Nelson's published account of an Ordovician ash bed in Tennessee. Nelson's hypothesis that the ash bed should be present throughout the "Lowville Sea" as far away as the northwest corner of Wisconsin was soon confirmed by Sardeson (1924) with the discovery of a K-bentonite bed at the Platteville-Decorah contact in Minnesota and Wisconsin. Sardeson was convinced that this was the same bed described by Nelson (1921). He subsequently discovered two other beds, a questionable K-bentonite in the upper part of the St. Peter Sandstone and a second bed in the Decorah Formation (Sardeson 1926a, 1926b, 1927, 1928 and 1934). Soon after, the St. Peter bed was shown not to be of volcanic origin, but the Decorah bed was found to contain the volcanogenic minerals biotite, apatite, and zircon (Allen, 1929).

By the late 1920s, Ordovician K-bentonites had been reported from many localities in the eastern United States. Nelson (1925, 1926) noted that there is more than one K-bentonite in the Middle Ordovician of Tennessee and Virginia. Multiple beds were reported by Butts (1926) in Alabama and by Stose and Jonas (1927) in Pennsylvania. In addition, six traceable K-bentonite beds were reported in numerous quarries in the Middle Ordovician succession of central Pennsylvania (Bonine and Honess, 1929). Giles (1927) noted the span of geologic time represented by the known K-bentonite beds and challenged the commonly held view that any clay bed in the upper Middle Ordovician originated from Nelson's (1921) volcano.

Investigations of the mineralogic and physical properties led Ross and Shannon (1925) to define bentonite as "a rock composed essentially of a crystalline clay-like mineral formed

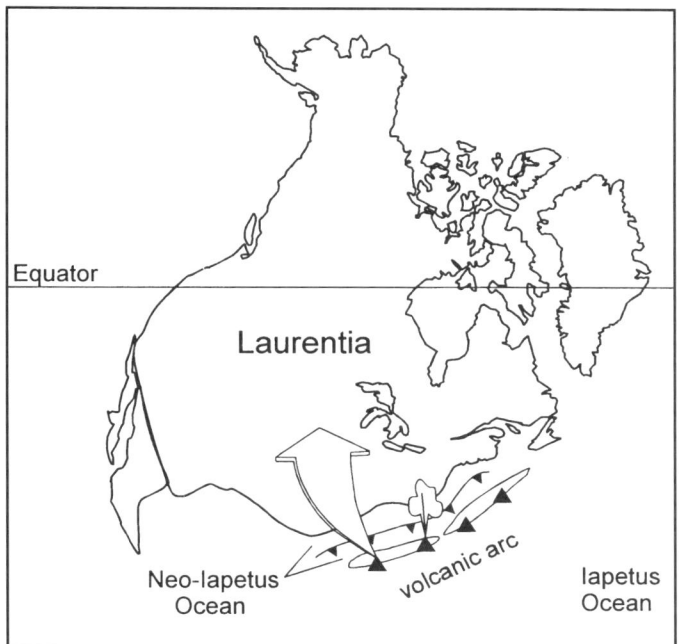

Figure 3. Middle Ordovician paleogeographic setting (modified from McKerrow et al., 1991). Volcanic ash from eruptions associated with an arc/continent collision was carried by southeast tradewinds into the interior of Laurentia where it fell into widespread cratonic seas.

by the devitrification and accompanying chemical alteration of a glassy igneous material, usually a tuff or volcanic ash" (Ross and Shannon, 1926, p. 79). Criteria for recognizing altered volcanic material was shown by Ross (1928) to include relict glass structure, the nature and habit of crystal grains and the physical properties of the clay minerals. In addition, Ross noted the occurrence of five K-bentonite beds in the Stones River (Carters Limestone) and overlying Trenton (Nashville Group) in Tennessee.

One of the first attempts to correlate K-bentonites on an interregional basis was made by Kay (1931) who traced the "Hounsfield metabentonite" from its type section near Watertown in upstate New York westward to southeastern Minnesota, a distance of 1,200 km. Kay further speculated that the bed could be correlated from eastern Missouri to Kentucky and Tennessee and that the progressive thickening of the bed toward the southeast implied a southern source. The notion that a single K-bentonite could be correlated over such a wide region was challenged by Whitcomb (1932) who pointed out that at least six beds are known from the same stratigraphic interval in Pennsylvania and that such a precise correlation as that by Kay (1931) is questionable. Furthermore, Whitcomb challenged Nelson's (1921) hypothesis that the K-bentonites originated from a single source or group of volcanoes in one small region. Rather, Whitcomb postulated that the Ordovician K-bentonites originated from a series of volcanoes, implying the volcanoes were geographically extensive.

In December 1933, a special symposium on K-bentonites was convened at the 46th Annual Meeting of the Geological Society of America in Chicago. Papers were presented by Ross and Kerr (1934) on bentonite and related clays, Rosenkrans (1934b) on problems involved in bentonite studies, Kay (1934) on the stratigraphic and geographic distribution of Ordovician altered volcanic materials and related clays, and Whitcomb (1934) on possible sources of Ordovician K-bentonites. In spite of a failed attempt to regionally correlate the "Hounsfield metabentonite," Kay continued to emphasize the potential utility of K-bentonites in long-distance correlations. Kay (1935) showed the known K-bentonites to range from Chazyan to mid-Mohawkian in age and to be widely distributed in eastern North America. Occurrences were shown in a series of stratigraphic columns of sections from New York, Ontario, Vermont, Virginia, Pennsylvania, Alabama, Michigan, Iowa, Wisconsin, Kentucky, and Tennessee. Although not explicitly stated, Kay implied that certain Middle Ordovician K-bentonite complexes extended over wide regions of eastern North America. On the basis of the progressive increase in thickness of most K-bentonites toward the southeastern United States, Kay suggested the source volcanoes were situated in western North Carolina. He illustrated this in a series of artfully constructed paleogeographic maps. The volcanic source of Ordovician K-bentonites was also investigated by Whitcomb (1935). He examined the areal extent of K-bentonites, the number of known beds, the zones of tectonic weakness which might have given rise to volcanoes, and the localities of known Ordovician volcanism in eastern North America. From this, he concluded that the volcanoes were situated in a "weak line" along the western edge of Appalachia, and that the remnants of the volcanic belt were probably buried beneath the Coastal Plain sediments. Nevertheless, he also concluded that local vents may have been present in the Mississippi Valley region. Unaware of plate tectonics and changing climatic patterns through time, Whitcomb noted that the present prevailing westerly winds would have carried most of the volcanic ash east of the Appalachians. He suggested that cyclonic storms were the probable mechanism that transported wind-borne ash westward deep into the interior of the continent.

Mid-1930s through late 1970s

With growing awareness of Ordovician K-bentonites, reports of their occurrence continued to show up in local stratigraphic studies. Ordovician K-bentonites have been observed in outcrops in northern Virginia (Rosenkrans, 1933), central and south-central Pennsylvania (Rosenkrans, 1934a; Whitcomb and Rosenkrans, 1934; Thompson, 1963), southwestern Virginia (Rosenkrans, 1936; Bates, 1939; Huffman, 1945; Miller and Brosgé, 1954; Miller and Fuller, 1954; Hergenroder, 1966, 1973; Haynes, 1992), New York (Kay, 1935, 1937; Walker, 1973; Cisne and Rabe, 1978; Cisne et al., 1982; Cisne and Chandlee, 1982; Goldman et al., 1994; Mitchell et al., 1994), central Kentucky (Young, 1940; Conkin, 1991; Conkin and Conkin, 1983, 1992;

Conkin and Dasari, 1986), eastern Tennessee (Fox and Grant, 1944; Rodgers, 1953; Coker, 1962; Milici, 1969), northwestern Georgia (Milici and Smith, 1969), central Tennessee (Wilson, 1949), northeastern Alabama (Drahovzal and Neathery, 1971), Upper Mississippi Valley (Mossler and Hayes, 1966), Ontario (Liberty, 1969; Trevail, 1990), and Quebec (Brun and Chagnon, 1979). Ordovician K-bentonites were also reported from the subsurface as far west as southwestern Texas (King, 1937) and eastern Kansas (Taylor, 1947).

One of the most extensive regional studies was published by Templeton and Willman (1963). By comparing detailed outcrop descriptions from numerous localities in eastern North America, they devised a K-bentonite stratigraphy for the Middle Ordovician (Champlainian Series) succession. They believed their predominantly lithologic correlation method was effective because of the great uniformity of sequences, gradual changes in facies, and the continuity of many distinctive units. Furthermore, they suggested that there are a few very widespread K-bentonites in the eastern midcontinent and many more localized beds in the Appalachians. Templeton and Willman agreed with Kay (1935) that the source volcanoes were situated in the area of North Carolina. Many of the K-bentonite beds in the Upper Mississippi Valley were formally named by Willman and Kolata (1978). Using conodont microfossils, Fetzer (1973) determined the relative biostratigraphic position of many Middle Ordovician K-bentonite beds of the Appalachians.

1980s to present

In recent years, there has been a growing trend toward the use of geochemical methods to determine the genesis and distribution of K-bentonite beds. The development of sophisticated, high-precision instrumentation to determine elemental concentrations has helped to characterize individual beds chemically and to determine age and paleogeographic, tectonic, and magmatic settings of source volcanoes.

Interest in Ordovician K-bentonites also has increased in recent years because of the success in identifying individual beds over wide regions. Six separate K-bentonite beds in the Middle Ordovician rocks of southwestern Ohio and northern Kentucky were correlated on the basis of their distinct chemical fingerprints (Huff, 1983). From an analysis of whole-rock samples, the chemical signatures of the Deicke and Millbrig K-bentonite Beds, the two thickest and most widespread K-bentonites in the Mississippi Valley, were identified in outcrop and subsurface from southern Minnesota to southeastern Missouri, a distance of about 900 km (Kolata et al., 1986, 1987). Using similar methods, five K-bentonites were chemically correlated in the Middle Ordovician Salona Formation in central Pennsylvania (Cullen-Lollis and Huff, 1986). Rare-earth element concentrations in primary apatite phenocrysts were used by Samson et al. (1988) to correlate Middle Ordovician K-bentonites from the Mississippi Valley to central Tennessee. The relative concentrations of certain phenocrysts, including labradorite, various Fe-Ti minerals, andesine, quartz, and biotite, were shown by Haynes (1992, 1994) to be useful in distinguishing between the Deicke and Millbrig K-bentonite Beds. Haynes (1994) also showed that east and southeast of the Cincinnati Arch the bulk mineralogy and chemistry of the beds change with increasing proportions of primary and authigenic non-clay minerals. Studies of unaltered volcanic ashes demonstrate that this is a common pattern in wind-dispersed deposits (Borchardt et al., 1971). Phenocryst size also increases toward the southeast, which suggests the source volcanoes were located east of the present Valley and Ridge Province (Zhang and Huff, 1994). Recently, Delano et al. (1994) demonstrated that a high degree of precision in chemical characterization of Ordovician K-bentonites can be obtained from fresh rhyolitic glass found as melt inclusions in phenocrysts of quartz, zircon, and apatite. The method has been effective in correlating K-bentonites in the Middle Ordovician of the Mohawk Valley, New York (Mitchell et al., 1994).

Ordovician K-bentonites contain phenocrysts that have been dated by various isotopic methods. Advances in dating technology over the past 30 years have led steadily to more reproducible and reliable ages. An average age of 419 ± 5 Ma (based on old constants) was obtained by conventional K-Ar from biotite in a K-bentonite collected near the top of the Stones River Formation (probable Millbrig K-bentonite Bed) of Alabama (Faul and Thomas, 1959; Faul, 1960). The Rb-Sr ages of biotite from beds in the Carters and Eggleston Limestones (probable Millbrig K-bentonite) of Tennessee range from 437 ± 50 Ma to 466 ± 50 Ma (Faul, 1960). The K-Ar ages for biotites and a sanidine from the Carters Limestone and Little Oak Limestone in Alabama range from 424 Ma to 453 Ma (Ghosh, 1972). Presently, K/Ar and Rb-Sr techniques are generally thought to be less reliable for isotopically dating Ordovician K-bentonites because of the difficulty in obtaining reproducible results. In both techniques, there is a propensity for a loss or gain of parent and daughter nuclear material. Zircons from K-bentonites in the Carters Limestone and Bays Formation of Tennessee yielded U-Pb ages ranging from 438 ± 10 Ma to 453 ± 10 Ma (Adams et al., 1960). Fission-track ages for three zircon samples from beds in the Tyrone Limestone (probable Deicke K-bentonite) of Kentucky were 447 ± 15 Ma, 462 ± 15 Ma, and 438 ±15 Ma (Ross et al., 1981). In addition, Ross et al. (1982) reported zircon fission-track ages of 454 ± 12 Ma, 456 ± 12 Ma, and 450 ± 10 Ma for K-bentonites from the Carters Limestone of Tennessee and the Decorah Formation (probable Millbrig K-bentonite) and Plattin Limestone (probable Deicke K-bentonite) of eastern Missouri, respectively. Isotopes of Nd and Sr from apatite gave an age of 457.1 ± 1.0 Ma for the Deicke K-bentonite Bed from eastern Missouri and central Tennessee (Samson et al., 1989). These authors suggest that the Nd and Sr isotopic data and U-Pb age data are consistent with the generation of the volcanic ash by anatexis of evolved continental crust.

Presently, the most widely cited ages for Ordovician

K-bentonites are from $^{40}Ar/^{39}Ar$ and U-Pb techniques. The $^{40}Ar/^{39}Ar$ age spectrum dates of biotite from the K-bentonites in the Stones River in Alabama and Carters Limestone of Tennessee ranged from 453 Ma to 456 Ma (Kunk and Sutter, 1984). Using similar methods, Kunk et al. (1985) obtained an age of 454.1 ± 2.1 Ma for a K-bentonite near the top of the Tyrone Limestone (probable Millbrig K-bentonite) in north-central Kentucky. Microgram-size zircon fractions from a K-bentonite in the Spechts Ferry Shale in Missouri (Millbrig K-bentonite) yielded U-Pb ages of 454.6 ± 1.6 Ma (Tucker et al., 1990). Both techniques suggest an age of approximately 454 Ma for the Millbrig K-bentonite.

Trace-element data obtained from the Millbrig K-bentonite Bed show a strong affinity with granites derived from volcanic arcs, indicating that the volcanic ash was generated along a continental-plate margin during collision (Huff et al., 1992). These authors calculate ash volume for the Millbrig and its probable counterpart in northwestern Europe to be on the order of 1,140 km^3 and that there is no observable extinction event in the major fossil invertebrates groups. In contrast, Sloan (1987) suggested that the widespread Deicke K-bentonite Bed in eastern North America forms a sharp break between Turinian and Chatfieldian shelly faunas.

MINERALOGICAL AND CHEMICAL CHARACTERISTICS

One of the most distinctive mineralogical features of Ordovician K-bentonite beds is their high proportion of clay minerals. Clay minerals and clay-size particles account for between 80% and 95% of the total volume of the beds, with the remaining portion made up of a variety of primary and secondary mineral grains, occasional rock fragments, and detrital grains. The clay minerals are the result of the devitrification of glass of primarily rhyolite to dacite composition that formed the bulk of the explosively erupted volcanic ash. Accompanying volcanogenic mineral grains include biotite, feldspar, quartz, apatite, zircon, leucoxene, or anatase, plus sulfide and carbonate minerals. Delano et al. (1990) reported garnet in K-bentonites from New York State and concluded that it represented transported source material from the lower crust. Yost et al. (1994) reported amphiboles in several Swedish K-bentonite beds and concluded on the basis of major element ratios, that they were not derived directly from parental magma. J. W. Delano (1995, personal communication) has also found amphiboles in Ordovician K-bentonites but concluded that their source was unclear. The origin of garnet and amphibole in K-bentonites remains problematic and requires further study. In some beds, particularly those close to the region of the source vents, the non-clay minerals frequently occur in alternating coarse- and fine-textured zones within the beds. The lateral continuity of these bands suggests that they are related to primary emplacement processes rather than postdepositional reworking.

Both primary phenocryst and whole rock chemical analyses have provided much useful information concerning original magmatic composition and have further contributed to interpretations of the tectonic settings of source volcanoes (Delano et al., 1990; Roberts and Merriman, 1990; Huff et al., 1993). Detailed studies of immobile large-ion lithophile (LIL) elements and rare-earth elements (REE) indicate most source volcanoes were associated with collision-margin tectonism and that the parental magmas were silicic in nature. These same elements have been found to be uniformly distributed within the K-bentonite beds and to have relative abundances unique to each bed. Thus, where biostratigraphic and lithostratigraphic evidence indicates the lateral equivalence of groups of K-bentonites, chemical fingerprinting can be used to correlate individual beds. Correlations of individual beds based on whole-rock immobile-element chemistry was first demonstrated by Huff (1983) and Kolata et al. (1986). Recently, Delano et al. (1994) have shown that microprobe analyses of melt inclusions in primary quartz phenocrysts can also provide distinctive fingerprints that are stratigraphically persistent on a local scale. The next section is a summary of the progress of research in each of these areas.

CLAY MINERALOGY

Ordovician K-bentonites were described in early reports by Nelson (1922), Ross (1928), and Kay (1935) as metabentonites based on their general similarity to younger Cretaceous and Tertiary bentonites of the western U.S. The Ordovician K-bentonites were presumed to have undergone low-grade metamorphism because of the lack of swelling clays. Weaver (1953) studied similar clay beds in central Pennsylvania and suggested they more correctly should be called K-bentonites since they had a high potassium content and lacked any obvious sign of metamorphism. Nelson (1922) interpreted the clays as resulting from the in situ alteration of explosively erupted volcanic ash contemporaneous with, or following shortly after, deposition in a marine environment. Subsequent work by Ross (1928) confirmed this interpretation. Ordovician K-bentonites consist primarily of mixed-layered illite/smectite (I/S) in various proportions of swelling and non-swelling components, with lesser amounts of chlorite, corrensite, and kaolinite (Brusewitz, 1986; Huff et al., 1988; Schumacher and Carlton, 1991; Haynes, 1992). This assemblage of clay minerals is characteristic of K-bentonites and generally distinguishes them from associated clay-bearing marine shales and siltstones whose components are largely detrital in origin. Kaolinite-rich altered ash beds, known as tonsteins, are commonly associated with coal measures throughout the Phanerozoic and provide an important source of information concerning the chemical behavior of volcanic glass during diagenesis (Bohor and Triplehorn, 1993).

The recognition of mixed-layered clays, particularly illite/smectite, by powder X-ray diffraction (XRD) techniques was greatly aided by the work of Weaver (1956) and subsequently by Reynolds and Hower (1970). These studies showed that it was possible to recognize the proportion of end-member components as well as the nature of order-disorder in I/S. The development of the computer program NEWMOD® (Reynolds, 1985) provided a useful and popular means for interpreting the nature of I/S interstratification in K-bentonites. Various graphic methods have also been developed and are widely used (Środoń 980; Watanabe, 1988). Recent work by Reynolds (1990, 1993) has extended computer modeling of XRD tracings to three-dimensional simulations of hkl (the generic Miller indices expression in crystallography) reflections in I/S. Reynolds has shown that disorder in R0 structures is mainly turbostratic, and that the illitic phases in some R1 and R > 1 structures are 1M (monoclinic) or 3T (triclinic) polytypes. The letter "R" refers to the Reichweite number or probability of nearest-neighbor ordering as defined by Jagodzinski (1949).

In the eastern Midcontinent of North America, the clay minerals in K-bentonites formed in a wide range of environments including shallow marine carbonate platform, carbonate slope facies, deep-water turbidites, black shale facies, and nearshore red-bed siliciclastics. Given the wide range of depositional environments, it is remarkable to see how consistent and pervasive the I/S-dominated mineralogy is. Secondary effects have been imposed by burial diagenesis and brine-flushing. As a result, the I/S found throughout the Mississippi Valley and eastward to the Cincinnati Arch is generally characterized by short-range ordering R1 to R2.5 with the illite component ranging from 62% to 85% (Kolata et al., 1986; Huff and Turkmenoglu, 1981; Elliott and Aronson, 1993). For glycol-treated samples the presence of a single diffraction maximum near $5°2\theta$ is indicative of R0 ordering. The appearance of short-range R1 ordering is indicated by a superstructure peak between $2°-3°2\theta$, and R3 ordering is revealed by the presence of two closely spaced reflections near $9°2\theta$ (Reynolds, 1980). In the Appalachian Basin, burial metamorphism and cation-rich fluid migration accompanying tectonism combined to produce largely R2.5 to R3 I/S with 85% to 95% illite (Elliott and Aronson, 1987; Haynes, 1994). Representative XRD tracings of the $<2\mu m$ size fraction are shown in Figure 4. In those portions of the basin that have been subjected to prehnite-pumpellyite grade metamorphism, chlorite, corrensite, and mixed-layer chlorite/smectite (Fig. 5) commonly occur as accessory clay minerals with R3 ordered I/S (Krekeler and Huff, 1993).

AGE OF ILLITIZATION

The chemistry of I/S clays from Paleozoic K-bentonites in the United Kingdom was studied in detail by Środoń et al. (1986) who reported a strong positive correlation between the amount of fixed K and the percent illite in I/S for the range 10% to 40% smectite. Fixation of K is affected by the increased substitution of Al for Si in the tetrahedral sheet of the smectite component and the consequent increase in total layer charge. One well-documented model of I/S formation invokes the transformation of smectite to illite by increased layer charge and the consequent fixation of K (Hower et al., 1976). Subse-

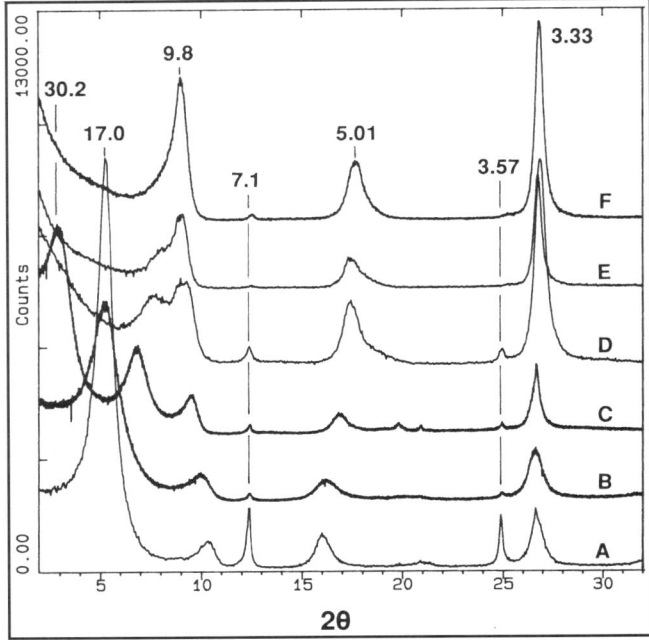

Figure 4. Selected powder X-ray diffraction (XRD) tracings of the $<2\mu m$ size fraction of Ordovician K-bentonites. Samples A through F are oriented and saturated with ethylene glycol and represent diverse geographic and stratigraphic localities. The sequence represents I/S dominated clay mineralogy ranging from random through long-range ordered (R) interstratification. Variable amounts of kaolinite are present as an accessory phase. Samples shown are (A) R0, I = 55%, (B) R0, I = 67%, (C) R1, I = 71%, (D) R2, I = 82%, (E) R2.5, I = 88%, and (F) R3, I = 95%.

quent studies (Altaner, 1989; Vasseur and Velde, 1993) have shown that K diffusion is controlled by kinetic factors during the change from smectite to illite. The transformation occurs as a multistage reaction involving three distinct crystalline phases. The first, R0 I/S, contains smectite from 100% to 50%, and transformation proceeds as a function of either time or temperature, or both. The second, described as a crystal growth process, contains R1 I/S with 50% to 5% smectite, and the third involves dissolution of I/S and crystallization of illite (Vasseur and Velde, 1993). Temperature, cation activity, and fluid migration appear to be critical factors in the formation of I/S, and many investigations into the kinetics of this process have centered on evidence from K-Ar age determination of the illite phase (Altaner, 1989). Illite has been shown to be an excellent K-Ar clock for all conditions below the blocking temperature of Ar (Aronson and Lee, 1986), and, for higher temperature regimes, K-Ar age determination can be used to constrain the onset of closure of the Ar system and related tectonic events (Huff et al., 1991).

In their study of Ordovician K-bentonites of Sweden, Velde and Brusewitz (1982) reported that, in general, thin beds tend to be uniformly rich in potassium, whereas thick beds tend to be K-rich at the edges and K-poor in the center, as a result of incomplete potassium metasomatism. The 1.8-m-thick Kin-

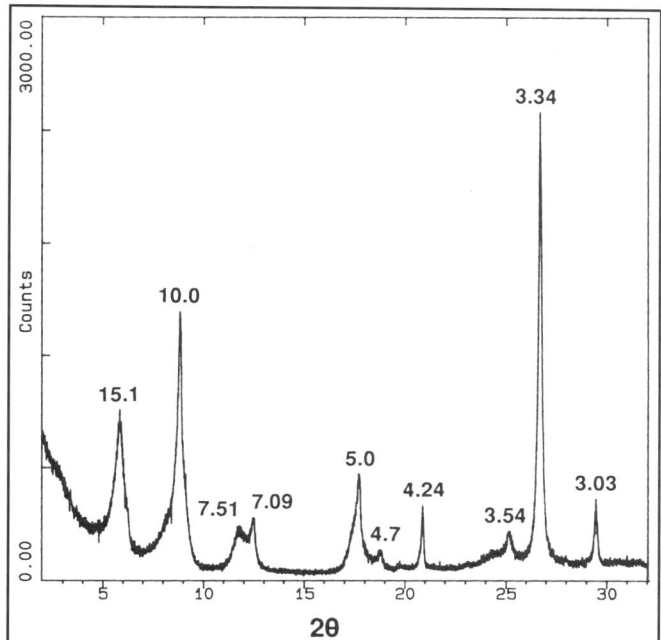

Figure 5. Powder X-ray diffraction tracing of the oriented and ethylene glycol-saturated XRD tracings <2µm fraction Millbrig K-bentonite from Harrisburg, Pennsylvania. The clay mineralogy is representative of K-bentonites subjected to low-grade metamorphic conditions. The I/S is R3 ordered with reflections at 10Å, 5Å, and 3.3Å and contains 98% illite. Kaolinite is indicated by reflections at 7Å and 3.5Å. Interstratified chlorite/smectite (C/S) is indicated by peaks at 15Å, 7.5Å, and 4.2Å.

nekulle K-bentonite Bed (Bergström et al., 1995) at Kinnekulle, Sweden, of Middle Ordovician age was studied in detail to provide further information on the timing and nature of illitization (Brusewitz, 1986, 1988). The depositional age of the Kinnekulle K-bentonite was measured at 455 Ma by the $^{40}Ar/^{39}Ar$ method on biotite (Kunk et al., 1984), whereas illite in the I/S gave K-Ar ages ranging from 336 Ma at the upper contact to 300 Ma in the center. The XRD data on the I/S showed 20% less smectite at the upper contact than in the center (Brusewitz, 1986). This variation appears to be systematic for the upper half of the bed, but not for the lower half, presumably because of the blocking effects of a thick basal chert zone. It would appear that the K-Ar ages do not record depositional or early diagenetic events associated with the devitrification of volcanic ash in sea water and the alteration to a smectite bentonite, but instead they reflect some later event during which K was mobilized in sufficient quantity to convert smectite to I/S. Permian diorite sills about 80 m above the Kinnekulle K-bentonite Bed were dated at 287 ± 15 Ma, and it is likely the heat from these dikes acted as a driving mechanism in mobilizing the fluids responsible for the K-enrichment reaction. The age of illitization of Middle Ordovician K-bentonites in the southern Appalachian Basin was studied in detail by Elliott and Aronson (1987, 1993). The percentage of illite in the I/S was found to be fairly constant with depth of burial and with presumed increase in burial temperatures. Although I/S ordering ranged from R > 1 to R3, it did not show a positive correlation with burial depth. Measured K-Ar ages from the illite component fell between 256 Ma and 313 Ma in the deepest parts of the basin, and these ages were found to generally agree with illitization ages of the Devonian Tioga K-bentonite in the central Appalachians. Samples of K-bentonite from the Cincinnati Arch on the western margin of the basin gave slightly younger ages (Elliott and Aronson, 1993). This evidence was used to support the hypothesis that a heated saline solution migrated from the more deeply buried proximal foreland basin during the Alleghanian orogeny as a result of convergent tectonism. A reversal of this age trend west of the Cincinnati Arch in the Illinois Basin was interpreted as a reflection of a separate brine circulation regime, similar to the one responsible for K-metasomatism elsewhere in the Illinois Basin (Hay et al., 1988).

The interpretation of K-Ar ages of illite is clearly not a straightforward matter. Their geological significance hinges upon their relationship to one or more thermal events in the history of the enclosing strata. Previous studies have shown that the Ar system can remain open in some minerals at temperatures in the range of low-grade metamorphic reactions (Dodson, 1973). The closure or blocking temperature of any radiometric decay scheme is that temperature below which no significant loss of the daughter product occurs; this is a function of several variables including cooling rate, particle size, and degree of crystal perfection of the host mineral. Although it has not been determined precisely for illite, the closure temperatures reported for some low-grade metamorphic micas range between 225 °C for biotite (Dodson, 1979) to 350 °C for phengitic mica (Hunziker, 1979).

The significance of measured ages diminishes in systems in which Ar diffusion has continued for very long time intervals, such as those approaching the half-life of the parent. Previous studies have shown that neoformed illitic clay minerals are excellent K-Ar clocks under sedimentary and diagenetic environmental conditions (Aronson and Lee, 1986). At higher temperatures, Ar diffusion may cause resetting of the ages to coincide with the blocking temperature during post-metamorphic cooling (Hunziker, 1986). The application of K-Ar ages in geothermometry and illitization studies is therefore constrained by temperatures characteristic of low-grade metamorphic environments. At higher temperatures, the reliability of the age measurements will be strengthened by thermal history data and measurement in a second radiometric system, such as Rb-Sr (Hunziker, 1986).

Layer charge of I/S in K-bentonites

Mixed-layer illite/smectite has been widely investigated for its role in sedimentary diagenesis and low-grade metamorphic reactions, but its most common occurrence in sedimentary rocks is in association with other clay minerals. For that reason it is frequently difficult to isolate I/S for detailed chemical and mineralogical studies. In contrast, the clay mineral composition of

Ordovician K-bentonites generally consists of I/S as the dominant clay material, often to the exclusion of all other clays. Therefore, K-bentonites have become the focus of research into questions that bear on the nature and origin of I/S in general. An illustration of this is provided by the numerous uses of K-bentonite I/S to investigate the nature and distribution of layer charge between the swelling and non-swelling components. Layer charge varies between end-member illite and smectite, and thus can provide key information about cation substitution in the octahedral and tetrahedral sheets which, in turn, can lead to improved understanding about the origin of I/S.

The structure of I/S in K-bentonites has been variously described according to several models that seek to account for the apparent coexistence of two distinct phases. The MacEwan crystallite model (Reynolds, 1980) describes I/S as composed of silicate layers about 10Å thick that are separated by nonexpanding illite and expanding smectite interlayers. The interlayers are arranged in the c-axis direction to form stacked crystallites with short-range and long-range order. Based on XRD data these crystallites consist of 5 to 15 silicate layers for a total of 50Å to 150Å in thickness. When viewed in transmission electron microscopy (TEM) images, however, many of these particles appear much thinner, which prompted Nadeau and co-workers (Nadeau, 1985; Nadeau et al., 1985; Nadeau and Bain, 1986) to propose the fundamental particle model. In this model, I/S is composed of silicate particles 10Å to 50Å thick that can hydrate and expand at their interfaces, thereby giving rise to interparticle diffraction characteristics that are indistinguishable from smectite. Both models have been extensively studied by computer simulation (Altaner and Bethke, 1989) and high resolution transmission electron microscopy (HRTEM) (Ahn and Buseck, 1990; Veblen et al., 1990). These studies show the presence of both expanding and nonexpanding silicate layers in the I/S structure but lack definitive information about the chemical composition of individual layers. Although the argument in favor of MacEwan crystallites was strengthened, modeling and HRTEM observation were unable to identify specific layer charges and compositions.

Layer charges of monomineralic clays can be measured by direct adsorption methods, by estimation from structural formulas, or from TEM observations. Środoń et al. (1992) investigated the apparent discrepancy between XRD and TEM data and found good agreement between the percent smectite as measured by HRTEM and the amount of fixed cations, particularly in samples with relatively small amounts of smectite. For samples with large amounts of smectite, the data are more scattered as a result of random measurement error. Nevertheless, their study highlighted the importance of using correct values of percent smectite for calculating smectite interlayer charge by instrumental methods.

The alkylammonium ion-exchange method offers the kind of probe that is sensitive enough to distinguish between the layer charges of illite and smectite (Lagaly and Weiss, 1969) and has been shown in numerous studies to not only discriminate between illite, smectite, and vermiculite, but also to track the heterogeneity of the charge distribution on the interlayer surfaces of the swelling clays (Lagaly, 1979). This is a method that utilizes the quantitative relationships between the layer charge and the area occupied by alkylammonium ions of varying carbon-chain lengths. Cetin and Huff (1995) studied a suite of Ordovician and Silurian K-bentonite clays with I/S interstratification ranging from R0 to R3 and containing between 30% and 97% illite layers. Since non-swelling clays such as illite are not responsive to short-term exchange reactions, only the swelling layers were examined. The orientation of alkylammonium ions in the interlayer space of swelling clays is a function of two factors, the layer-charge density of the host clay and the length of the alkylammonium chloride chain. The latter varies with the number of carbon atoms (N_c) appended to the alkyl complex to form the various alkylammonium derivatives. As successively longer chains (increasing N_c) are exchanged into the interlayer space of smectites and low-charge vermiculites, the orientation of the chains will change from a monolayer to a bilayer configuration (Fig. 6). If the mineral has a homogeneous charge distribution, the transition occurs as a sharp jump from about 13.6Å to about 17.7Å between two successive values of N_c (Fig. 6A). However, Lagaly and Weiss (1969) have shown that most smectites and low-charge vermiculites have heterogeneous rather than homogeneous layer charge distribution patterns, and thus the transition from monolayer to bilayer will occur over a range of N_c values greater than two (Fig. 6B) since both arrangements are co-existing. The resulting XRD tracing is a series of non-integral basal spacings between 13.6Å and 17.7Å.

The I/S clays in Ordovician K-bentonites with >15% swelling layers commonly have heterogeneous layer charge densities ranging between 0.32 and 0.45 equivalents per half-cell formula unit, as is commonly found in smectites (Lagaly and Weiss, 1976). Ordovician K-bentonites with <15% swelling layers show a continuous and more or less linear relationship between the first order d-spacing (spacing between the planes of atoms in a mineral structure) of the expandable component and N_c.

A significant conclusion from these data is that expandable interlayers in I/S from Ordovician K-bentonites with >15% swelling retain a smectitic character regardless of the proportion of illite, even after the particles have been disarticulated in suspension and resedimented on a laboratory substrate (Cetin and Huff, 1995). Moreover, there is not a close correlation between the magnitude of the interlayer charge and the proportion of swelling layers as determined by XRD. For I/S with <15% swelling layers, the similarity of charge densities estimated by the alkylammonium ion method and those calculated from structural formulas indicates the presence of a vermiculite-like component. Therefore, it may be concluded that the expandable interlayers are more properly described as vermiculite and that the interstratified phase might be more properly described as illite/vermiculite (Cetin and Huff, 1995). These data suggest that two, and possibly three, mineralogic phases co-exist in I/S from K-bentonites and that their wide range of

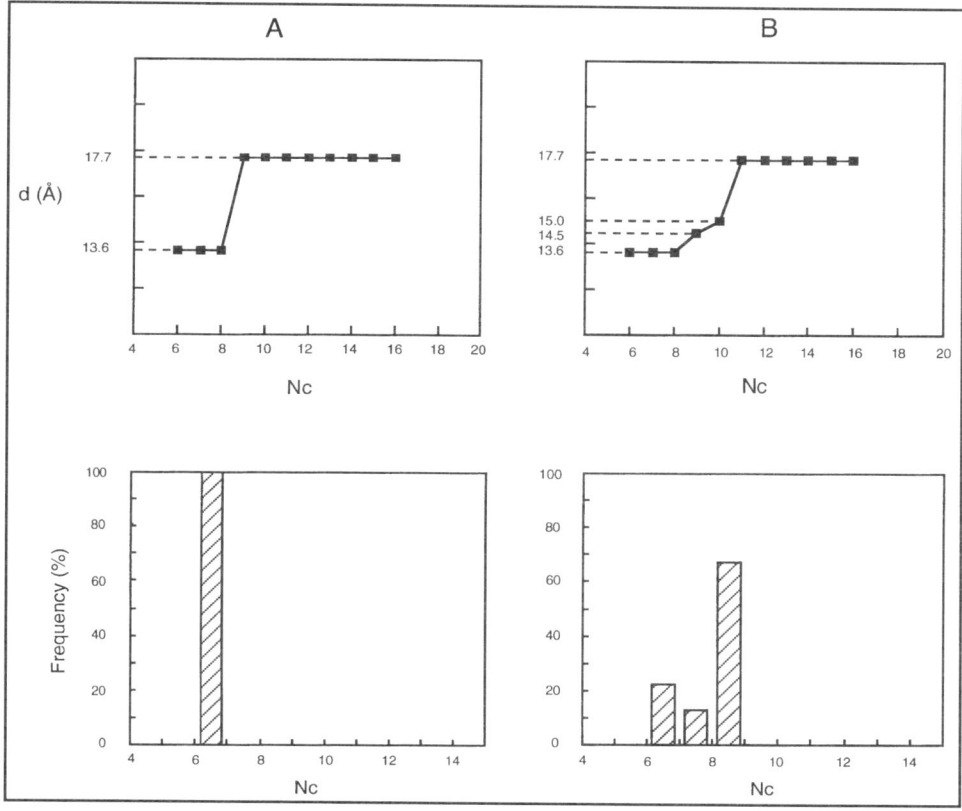

Figure 6. Behavior of n-alkylammonium treated I/S from the work of Cetin and Huff (1995), with Nc representing the number of carbon atoms in alkyl chain. The short-term exchange reactions of alkylammonium ions affects the smectite phase only, and the structural response to increasingly longer chain length reflects the distribution pattern of interlayer charge sites. (A) Alkylammonium ion exchange in the smectite component of K-bentonite I/S in which the abrupt transition from monolayer (13.6Å) to bilayer (17.7Å) configuration indicates a homogeneous charge distribution. The histogram represents interlayer cation density as a function of the length of the carbon chain (Nc). (B) The transition from monolayer to bilayer arrangement occurs over a range from Nc = 8 to Nc = 11, indicating a heterogeneous layer charge. Cation density as a function of Nc is shown below.

apparent stabilities is consistent with a mechanism of layer transformation of smectite to illite, rather than dissolution and reprecipitation, as has been argued elsewhere (Nadeau et al., 1985). These layer-charge data do not preclude the existence of interparticle diffraction, but rather they establish the smectitic character of what have been described as illitic particle surfaces in I/S with >15% expandability. A layer-by-layer transformation of I/S, in which the interlayer charge characteristics of a precursor smectite are inherited appears more likely than a neoformation mechanism, since neoformation requires precipitation of illite particles with surface charges significantly higher than that of smectite.

High resolution transmission electron microscopy

Illite/smectite in Ordovician and Silurian K-bentonites has been studied by high resolution transmission electron microscopy (HRTEM) in both the semi-dispersed, untreated state, and following alkylammonium ion exchange (Cetin and Huff, 1995). Samples with R > 1 ordering are dominated by crystallites with 20, 30, and 40 lattice fringes, and these were interpreted as representing R1, R2, and R3 ordering, respectively, in which the heavier dark fringes correspond to smectite layers. This interpretation was supported by HRTEM images of an R > 1 sample saturated with hexadecylammonium chloride in which the expanded interlayers within the crystallites were constructed mainly of 20Å, 30Å, and 40Å thick units. Previous interpretations of similar stacking arrangements in other I/S clays attributed these dimensions to polytypic varieties (Iijima and Busek, 1978). Alternative interpretations include the isolated sub-units as the result of stacked illitic layers of fundamental particles (Nadeau et al., 1984; Nadeau, 1985). Discrete particles of similar dimensions may give rise to interparticle diffraction effects that have the same appearance as interstratified I/S. However, in the fundamental particle model there is no prescribed crystallographic continuity between adjacent particles, and thus HRTEM lattice fringes should not show cross-fringes. Cetin and Huff (1995) show that 4.5Å cross-fringe images do occur in R > 1 ordered samples indicating that the range of coherent stacking exceeds the size of the fundamental particles. This observation is consistent with recent studies by Ahn and Buseck (1990) and Veblen et al. (1990) who also indicated that the range of coherent stacking exceeds the size of fundamental particles in I/S.

Petrography

Ordovician K-bentonites contain differing amounts of primary and secondary non-clay minerals, some of which provide additional stratigraphic and tectonomagmatic information. The main primary minerals, mostly in the form of isolated, euhedral phenocrysts, are quartz, biotite, plagioclase and potassium feldspar, ilmenite, apatite, zircon, and magnetite. The principal authigenic minerals are potassium feldspar, albite, TiO_2 as rutile and anatase, hematite, pyrite, calcite, and gypsum. Garnet and amphibole have been reported and are thought to represent mixing of the ash with other source rocks. Haynes (1992, 1994) has described the characteristics of these minerals in some detail for the Deicke and Millbrig K-bentonites in the southern Appalachians. Where primary plagioclase has been preserved, the two K-bentonites can be distinguished from one another. The primary plagioclase in the Millbrig is andesine (An30–50), and the primary plagioclase in the Deicke is laboradorite (An50–70). Most feldspar grains in Ordovician K-bentonite are authigenic albite and K-feldspar. Occasional sanidine grains have been reported but are not common. Authigenic feldspars in the Deicke and Millbrig include albitized K-feldspar, secondary albite after plagioclase, and secondary K-feldspar after either sanidine or plagioclase. Subsequent studies of the Deicke and Millbrig by McVey (1993) reported the preservation of polysynthetic twinning in secondary albite. Thus twinning alone is an inaccurate guide to the origin of feldspar grains.

Delano et al. (1990) analyzed garnets in Ordovician K-bentonites from New York State and concluded that they represent the influence of a high-pressure regime, probably the lower crust, on the composition of the parent magma. They further concluded that this constrained the tectonic setting of the source volcanoes to, most likely, an active plate-margin setting.

The Deicke contains Fe and Fe-Ti phases which account for as much as 10% of the non-clay fraction in some samples (Haynes, 1992, 1994). Pyrite is the most common opaque mineral, with primary magnetite and ilmenite less common. Anatase or rutile are the common pseudomorphic forms of ilmenite. Quartz is particularly abundant in the Millbrig and may account for as much as 25% of the non-clay fraction (Haynes, 1992, 1994). Deicke samples generally contain <1% quartz. Individual grains are subhedral to anhedral with many showing the remnants of pyramidal faces fractured during the eruption. Quartz grains host numerous melt inclusions whose chemistry provides important insight into the composition of the parental magma, as well as serving as a reliable source of stratigraphic components for chemical fingerprinting (Delano, 1992).

Euhedral to anhedral biotite constitutes up to 30% of the non-clay fraction in the Millbrig K-bentonite in the southern Appalachians and also serves as a reliable discriminator from the Deicke in hand sample. The Kinnekulle K-bentonite of northwestern Europe is similarly marked by a high content of biotite and is considered to be stratigraphically equivalent to the Millbrig (Huff et al., 1992). Biotite is also abundant in K-bentonite beds that lie stratigraphically above the Millbrig and Kinnekulle, but not in those lower in the section. In samples where biotite has been well preserved it serves as a reliable source of age dates based on the $^{40}Ar/^{39}Ar$ method (Kunk et al., 1984; Kunk and Sutter, 1984).

Both euhedral zircon and apatite crystals commonly form on the order of 1% of the non-clay fraction of most Ordovician K-bentonites. They average between 0.1 and 0.5 mm in length with length-to-width ratios of 3:1 to 8:1 for zircons and 2:1 to 3:1 for apatites. Zircons with high length-to-width ratios give more consistent U-Pb ages and tend to have fewer inherited cores than the shorter crystals (R. D. Tucker, 1994, personal communication). Samson et al. (1988) used apatite chemistry as a basis for stratigraphic correlation of Ordovician K-bentonites and showed the feasibility of using precise chemical analysis of single phases for stratigraphic purposes. Apatite crystals commonly include melt inclusions that may also provide information about the composition of parental magmas. Preliminary work indicates the glasses are rhyolitic in composition.

Chemistry

Chemical analyses of Ordovician K-bentonites provide additional information about their origin and stratigraphy. Large-ion lithophile elements, such as Th, can be similarly used to distinguish between ocean-ridge magmas and those associated with destructive plate-margins, particularly back-arc and marginal basin settings. The K-bentonite data clearly indicate origin of the volcanic ash from a Th-enriched, calc-alkaline, destructive plate-margin volcanic setting. Such rocks are characteristically rhyolites, trachytes, and trachyandesites representing evolved liquids that concentrate incompatible elements such as Th. Huff et al. (1993) showed that for Ordovician K-bentonites in Great Britain, Th-enrichment is greater for the Caradoc/Ashgill samples than for the Llanvirn/Llandeilo samples. This suggests a change through time to more evolved magmas erupted in a back-arc or plate margin extensional setting. In that study, 25 samples of Middle and Upper Ordovician K-bentonites from graptolitic shales representing *D. artus* through *D. anceps* biozones from localities in Wales, the Welsh Borderland, and the Southern Uplands of Scotland were analyzed for major and trace elements, and the data were plotted on tectonic and magmatic discrimination diagrams. Many of the immobile elements measured have bulk distribution coefficients <1 and are partitioned into the liquid phase during partial melting and into the evolving liquid of a crystallizing magma (Wood et al., 1979). They record differences in magmatic origin ranging from direct fractional crystallization of mantle-derived magma through varying degrees of contamination by partial melting of lower crustal felsic rocks and incorporation of subducted sediments, and thereby they might be expected to retain characteristic signatures of the products of Plinian eruptions from such sources. They may also reflect crustal melting and assimilation of crustal rocks at shallow depths in magma chambers as demonstrated by inherited zircons with Precambrian cores (Samson et al., 1988). Thus, the diagrams provide information on both tectonic setting and magmatic composition of the volcanism responsible for the K-bentonites. These data indicate a calc-alkaline plate–margin source, similar to those of the U.K. samples.

Chemical fingerprinting

Long-range stratigraphic correlation of K-bentonites based on unique element or element groupings was first employed by Huff (1983) and Kolata et al. (1986, 1987). Most of the detailed chemostratigraphic work has concentrated on four mid-Mohawkian beds including the Deicke, Millbrig, Elkport, and Dickeyville K-bentonites (Willman and Kolata, 1978) in the basal part of the Galena Group. These beds occur in numerous widely distributed outcrops where stratigraphic control is very good. Kolata et al. (1987) correlated the Deicke and Millbrig K-bentonite Beds by establishing that the beds had distinctive and characteristic trace-element compositions recognizable over the region from St. Paul, Minnesota, to southeastern Missouri. Differences in elemental abundance among the four beds were apparent in bivariate plots of a few of the elements. However, the plots do not always give clear, unequivocal clusters of data points for each bed, and it is awkward to assimilate information from several plots at the same time. Hence, a method of evaluating several of the elements together was needed to determine if a distinctive chemical signature for each bed could be defined. Discriminant function analysis, a multivariate statistical method, was employed to analyze the 26 variables (elements) and four groups (beds). The discriminant analysis method was selected because it seeks to statistically distinguish between two or more groups of samples using a set of variables that are thought to differ between groups (Klecka, 1981). The mathematical objective is to weight and linearly combine the discriminating variables so that the groups are forced to be as statistically distinct as possible. The discriminant function is especially useful in detecting subtle combinations of variables that, when considered together, result in a larger group difference than any of the variables considered alone. If that difference is significant, then the computed discriminant functions can be used to assign unknown samples to one of the groups.

When there are large numbers of variables of unequal or uncertain discriminating power, it is usually necessary to use a stepwise algorithm for selecting variables to include in the discriminant functions. The stepwise procedure begins by selecting the single best discriminating variable according to a separation criterion described below. The second variable selected is the one that, in combination with the first, best improves the value of the discrimination criterion. The third and subsequent variables are similarly selected according to their ability to contribute to further discrimination. An example of the application of discriminant analysis is shown in Figure 7, which compares samples from the Middle Ordovician Deicke, Millbrig, and V-7 (Rosenkrans, 1936) in the southern Appalachians. Chemical differences between the three beds are maximized by the analysis, and a discriminant diagram is constructed with boundaries between the groups defined by the midpoints between group means. The computer-generated discriminant scores can then be used to classify any unknown sample suspected of representing one of the three beds. Discriminant analysis thus offers a weight-of-evidence criterion that cannot be observed by measuring any of the individual variables in the model, and thereby discriminant analysis extends the limitations of conventional empirical analysis by maximizing the advantage of variable assemblages.

BIOSTRATIGRAPHIC FRAMEWORK

A necessary first step in correlating K-bentonites is to determine their relative positions within an established biostratigraphic framework. At present, conodonts provide greater biostratigraphic resolution than any other fossil group for the Ordovician platform carbonate rocks (Bergström, 1971; Sweet et al., 1971; Sweet and Bergström, 1976; Sweet, 1984), whereas graptolites are the most useful in the basinal facies that characterize the pericratonic regions of eastern North America (Finney, 1982; Goldman et al., 1994). Zonal ties with Lower and early Middle Ordovician trilobites and brachiopods also have been proposed by Ross et al. (1982); however, benthic shelly faunas

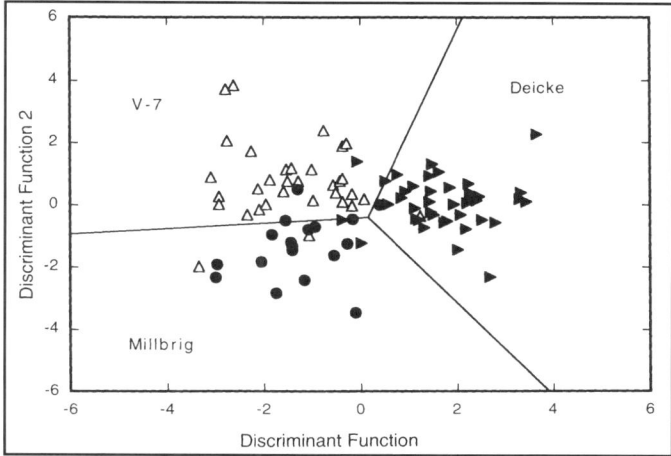

Figure 7. Discriminant function plot of Deicke, Millbrig, and V-7 K-bentonite samples from the southern Appalachians, based on the computed first and second discriminant functions. A hierarchical ranking of 10 immobile elements was produced by the discriminant analysis of the three beds. At the 95% confidence level, the three beds can be distinguished from one another with a high degree of certainty. Symbols for K-bentonite samples: solid circles, Millbrig; open triangles, V-7; solid triangles, Deicke.

in the eastern Midcontinent United States tend to be ecologically restricted.

The present state of Ordovician biostratigraphy, particularly for the Lower Ordovician, is somewhat unsettled because of disagreements on the significance and correlation of the various zonal schemes. A biostratigraphic analysis of the Ordovician is now the focus of several working groups of the Ordovician Subcommission of the International Commission on Stratigraphy. Consequently, there is no universally accepted subdivision of the Ordovician System even at the series level. In the absence of formal international agreement, the biostratigraphic terminology and British, Baltoscandian, and North American series and stage designations used here follow Bergström (1990) for the Middle and Upper Ordovician and Sweet and Bergström (1986) for the Lower Ordovician (Plate 1).

The Late Middle Ordovician standard stage classification follows Leslie and Bergström (1995b). Hence, in order to avoid confusion with the lithostratigraphic term Black River Formation or Group, strata formerly assigned to the Blackriveran Stage are placed in the Turinian Stage, named by Fisher (1977) for Turin Township in the Black River Valley, New York. The base of the Turinian coincides with the base of the *Baltoniodus gerdae* conodont Subzone (Fig. 8). The Chatfieldian Stage, overlying the Turinian, was proposed by Leslie and Bergström (1995b) to resolve ambiguity with the late Mohawkian stage-level nomenclature. The Chatfieldian is named for the fossiliferous strata of the Galena Group near Chatfield, Minnesota. The base of the Chatfieldian is marked by the widespread Millbrig K-bentonite Bed, and its upper boundary is the base of the superjacent Edenian Stage of the Cincinnatian Series. As pointed out by Leslie and Bergström (1995b), the base of the Chatfieldian appears to correspond to a level in the succession cut out by the unconformity between the Black River and Trenton Groups, which traditionally has been picked as the base of the "Rocklandian" and "Trentonian."

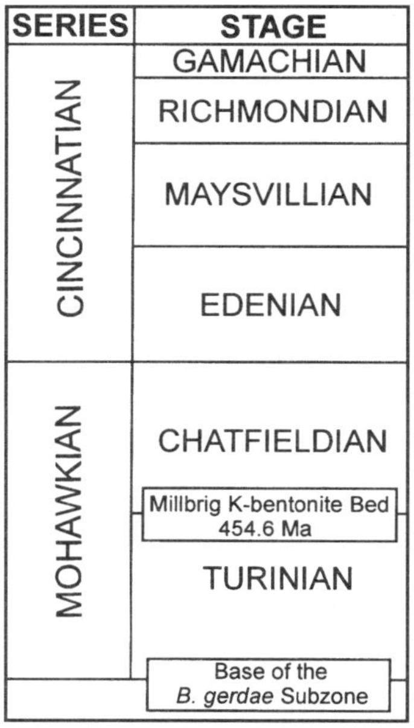

We have attempted herein to show the stratigraphic sequence of K-bentonites from the various regions of eastern North America in terms of the known faunal associations. In most cases, conodonts provide the greatest biostratigraphic resolution. For most localities discussed in this book, the biostratigraphic position of individual beds is extrapolated from the nearest studied section. As a result, in some regions the biostratigraphic control is well constrained, whereas in other places it is less precise.

In a biostratigraphic analysis of Middle Ordovician K-bentonites, Fetzer (1973) recognized 11 clusters or complexes of K-bentonite beds occurring in definable stratigraphic intervals. He correlated the complexes with North Atlantic and Midcontinent conodont zones and showed that the complexes range in age from the *Pygodus serra* Zone (*Eoplacognathus foliaceus* Subzone) through the *Amorphognathus superbus* Zone. Utilizing the Composite Standard Section (CSS) proposed by Sweet (1984, 1988), Bergström (1989) named and defined seven K-bentonite complexes in terms of CSS units. He pointed out that, in the case of sections not included in the CSS, traditional conodont biostratigraphy can be used to locate Sweet's CSS-defined zonal boundaries. Furthermore, Bergström (1989) noted the similarity in temporal distribution of K-bentonites in North America and Baltoscandia and suggested that there may be a common source of volcanic ash in the northern Iapetus Ocean. Recently, Leslie (1995) conducted an extensive biostratigraphic study of K-bentonites in the Deicke-Millbrig interval.

Spatial differentiation patterns of Middle Ordovician shelly, graptolite, and conodont faunas in eastern North America show an amphicratonic distribution characterized by three major facies (Jaanusson and Bergström, 1980). The facies are best exposed in parallel allochthonous thrust belts in the Appalachians. The innermost belt, Lee Confacies, coincides with the central interior of eastern North America and is characterized by a carbonate platform facies containing the Midcontinent conodont fauna (Plate 1). The outermost, Blount Confacies, consists primarily of siliciclastics containing Pacific Province type graptolites in association with dominantly North Atlantic Province conodont faunas. The central belt, Tazewell Confacies, consists of mixed carbonate and siliciclastic rocks containing a mixture of Blount and Lee faunas. In order to correlate K-bentonites regionally, it is necessary to know their temporal relations to the key conodont and graptolites zones.

K-BENTONITE DISTRIBUTION IN EASTERN NORTH AMERICA

The discussion in this book of Ordovician K-bentonites in eastern North America is organized by regions, with focus on the outcrop belts and cratonic basins. This approach is conve-

SERIES	STAGE
CINCINNATIAN	GAMACHIAN
	RICHMONDIAN
	MAYSVILLIAN
	EDENIAN
MOHAWKIAN	CHATFIELDIAN
	Millbrig K-bentonite Bed 454.6 Ma
	TURINIAN
	Base of the *B. gerdae* Subzone

Figure 8. Classification of late Middle and Late Ordovician Series and Stages for North America (Leslie and Bergström, 1995). Base of the Chatfieldian Stage is situated at the widespread Millbrig K-bentonite, and that of the Turinian Stage is situated at the base of the *B. gerdae* Subzone.

nient, particularly in reviewing the literature, because most previous studies concentrated on local outcrop and subsurface investigations. We review the stratigraphy of Ordovician K-bentonites in North America and seek to identify beds that are widespread and have event-stratigraphic significance. Our database consists of approximately 150 outcrop and 450 subsurface localities distributed over the eastern half of the United States and parts of eastern Canada (Fig. 9). Specific localities referred to in the text are listed in Appendix 1—Locality Register.

Historically, stratigraphic research of Ordovician K-bentonite beds in North America has been confined to local outcrop investigations. Only a few attempts have been made to correlate through the subsurface between outcrop belts (Kolata et al., 1986, 1987; Huff and Kolata, 1990). Wherever possible, we have attempted to trace beds through the subsurface on wireline logs and in cores and drill cuttings. We used self potential, resistivity, gamma ray, neutron, sonic, laterologs, and microcaliper logs. The K-bentonites are characterized by thin sharp peaks of natural radiation and water saturation and as thin caved zones on microcaliper logs. Generally, gamma-ray and neutron logs provide the best resolution of K-bentonite beds. The beds typically have high gamma ray counts, reflecting the relatively high potassium content. Neutron logs indirectly measure the porosity of rocks. Compared to the enclosing carbonate rocks, K-bentonites are relatively porous; thus they exhibit a deflection of the neutron curve in the direction of increased porosity. Subsurface recognition of K-bentonite beds is most successful where they occur within pure carbonate rocks, as found in the Middle Ordovician Trenton and Black River Groups and their lateral equivalents. Commonly, K-bentonite beds as thin as 5 cm produce a deflection on gamma-ray/neutron logs. Our wireline log correlations are based on an examination of hundreds of logs, and the cross sections were cross correlated with numerous tie lines. All wireline logs figured herein were traced from expanded scale (1 in = 20 ft) logs.

Upper Mississippi Valley

The outcrop belt includes Ordovician rocks exposed along the bluffs of the Mississippi River and its tributaries and on the crest of the Wisconsin Arch. Many quarries, roadcuts, and natural outcrops bearing Ordovician K-bentonite beds are present in southeastern Minnesota, southwestern Wisconsin, northeastern Iowa, and northwestern Illinois (Fig. 10). In addition, a

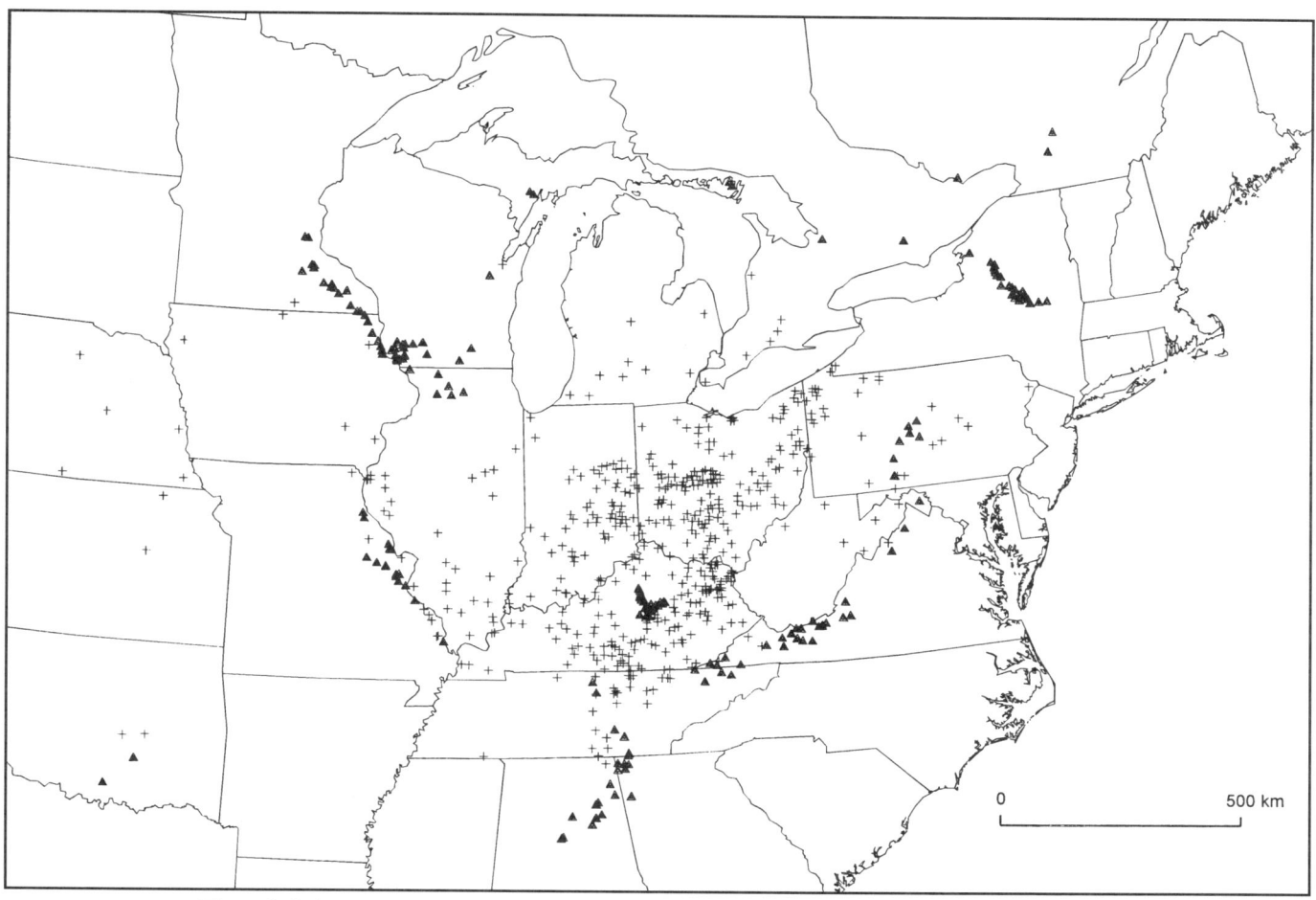

Figure 9. Point source data used in this investigation. Triangles refer to outcrops and pluses to drill holes from which wireline logs and/or subsurface samples were studied.

small number of cores and wireline logs are available from the subsurface at the respective state geological surveys. A review of literature on K-bentonite beds in this region was published by Templeton and Willman (1963), Mossler and Hayes (1966), Levorson and Gerk (1972), Willman and Kolata (1978), Kolata et al. (1986, 1987), and Levorson et al. (1987). The detailed stratigraphic descriptions prepared by Levorson and Gerk are an important source of unpublished information on K-bentonite occurrences in northeastern Iowa and southeastern Minnesota. Their descriptions are on file at the Illinois, Iowa, and Minnesota Geological Surveys.

A composite stratigraphic column for the Upper Mississippi Valley region is shown in Figure 11. The oldest known K-bentonite in the region occurs locally in outcrops at the base of the St. Peter Sandstone in Wisconsin (Woodard, 1972; Hay et al., 1988) and in the subsurface in northern Illinois (Templeton and Willman, 1963). At Hanover, Wisconsin (Appendix 1, locality 1), the bed is up to 30 cm thick and has been altered to brittle beds of pink to orange authigenic K-feldspar. Near Utica in northern Illinois (locality 2), several clay beds suspected of being K-bentonites have been described in outcrop (Cady, 1919). The age of deposition is uncertain because the bed(s) rests on the sub-Tippecanoe unconformity where a significant part of the Whiterockian Series is absent as a result of subaerial erosion. The bed(s) may represent the residuum of several ash falls that accumulated locally in depressions on the exposed carbonate rocks of the Sauk Sequence prior to the transgressive cycle that began with deposition of the St. Peter Sandstone. The fact that the bed(s) is preserved at all suggests that it was deposited during late Whiterockian time before removal by erosion.

Locally, a 1 cm to 2 cm thick K-bentonite is present at the top of the Mifflin Formation, Platteville Group, in northern Illinois (localities 3 and 26). Even where the bed is thin, its position is commonly marked by a prominent bedding plane (Willman and Kolata, 1978). We suspect that this may be the clay bed overlying the sponge and cystoid horizon reported by Ulrich and Everett (1890) and Kolata (1975) near Dixon, Illinois (locality 3).

The most widespread and persistent beds in the Upper Mississippi Valley region occur in the late Mohawkian and early Cincinnatian Galena Group (Fig. 11). Several beds have been formally named (Willman and Kolata, 1978) and traced widely in outcrop (Templeton and Willman, 1963). In contrast to recent reports (Templeton and Willman, 1963; Willman and Kolata, 1978; Kolata et al., 1986), we classify the Decorah as a formation divided, in ascending order, into the Castlewood Limestone (Missouri) or its equivalent, the Carimona Limestone (Minnesota), Spechts Ferry Shale, Guttenberg Limestone and Ion Shale Members. This scheme avoids the use of the "subgroup" level of classification, thus following the North American Stratigraphic Code (1983) but maintaining the geographic designation in the member name.

The Deicke K-bentonite Bed (I-1 bentonite of Mossler and Hayes, 1966), one of the most widespread K-bentonites in eastern North America, occurs near the base of the Decorah Formation. The bed can be traced in outcrop and in the subsurface on wireline logs and in cores (Plate 2) from its type section near St. Louis, Missouri (locality 4), northward into the Upper Mississippi Valley outcrop belt. In the type area, the Deicke overlies the pure limestone of the Platteville Group and is overlain by approximately 2 m of the Castlewood Limestone Member of the Decorah Formation. The Deicke produces a prominent and readily identifiable deflection on gamma-ray/neutron logs in the subsurface of eastern Missouri, western Illinois, and eastern Iowa. As a result, the Deicke can be traced with confidence into northeastern Iowa and the Upper Mississippi Valley outcrop belt (Fig. 10). It is also obvious from this correlation that the Castlewood Member is equivalent to the Carimona Limestone Member of the Decorah in Iowa and Minnesota (Plate 2). In the Upper Mississippi Valley region, the Deicke is generally 5 cm thick, but locally it is as much as 10 cm.

One of the earliest descriptions of the Deicke was made by Sardeson (1924, 1926a, 1926b, and 1927) from outcrops in southeastern Minnesota and southwestern Wisconsin (Figs. 12 and 13). Sardeson assumed that it was the same bed described by Nelson (1922) in parts of Kentucky and Tennessee, now confirmed by outcrop and subsurface correlations. Sardeson (1928, 1934) later described another bed, now recognized as the Millbrig K-bentonite, in the Decorah Shale at Minneapolis, Minnesota.

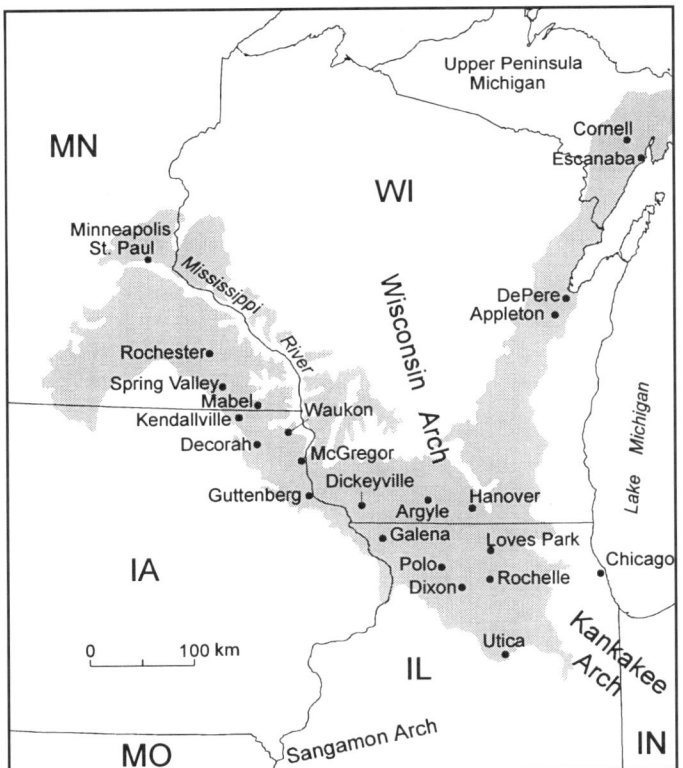

Figure 10. Upper Mississippi Valley region, showing major structural features and outcrop belt (shaded).

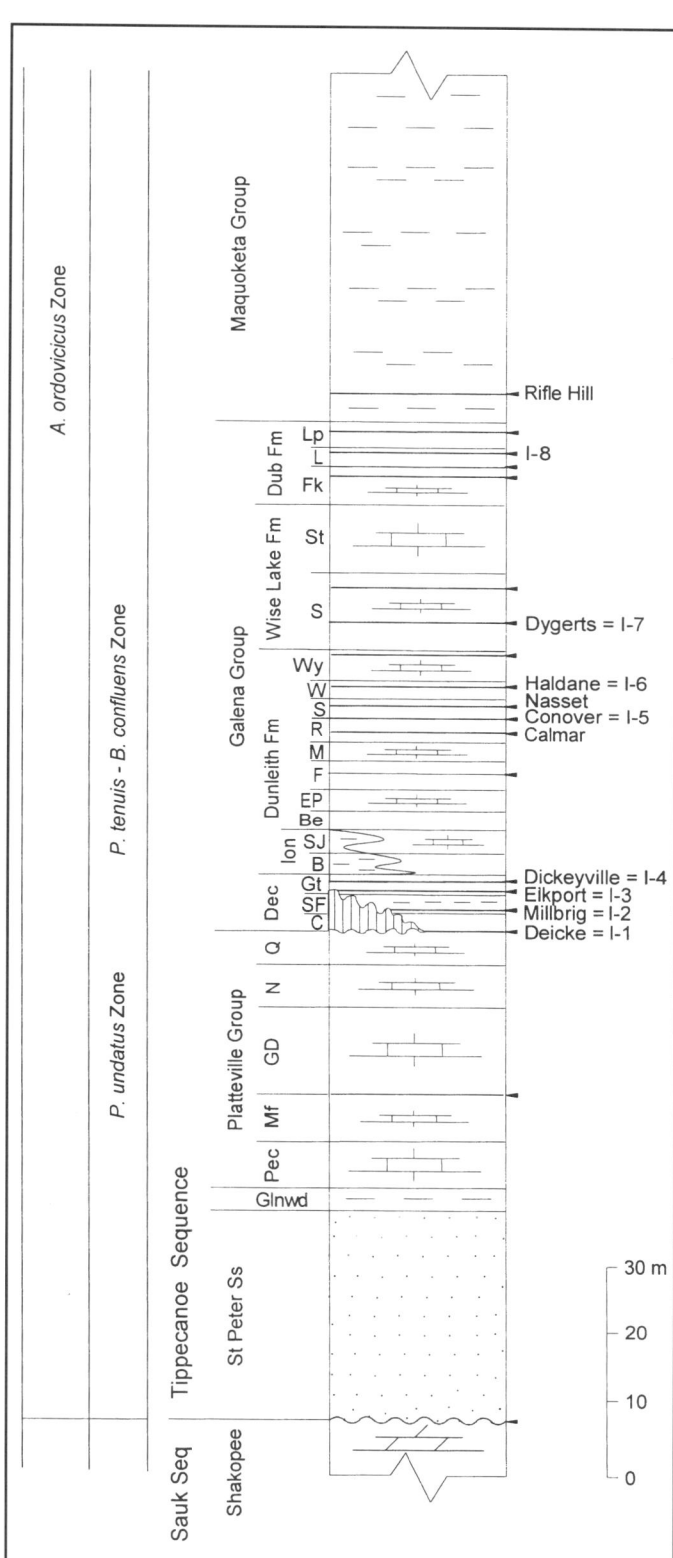

Figure 11. Generalized stratigraphic column of the Upper Mississippi Valley region. Abbreviations: Sandstone, Ss; Glenwood Formation, Glnwd; Pecatonica Formation, Pec; Mifflin Formation, Mf; Grand Detour Formation, GD; Nachusa Formation, N; Quimbys Mill, Q; Decorah Formation, Dec; Castlewood Limestone Member, C; Spechts Ferry Shale Member, SF; Guttenberg Limestone Member, Gt; Buckhorn Member, B; St. James Member, SJ; Beecher Member, Be; Eagle Point Member, EP; Fairplay Member, F; Mortimer Member, M; Rivoli Member, R; Sherwood Member, S; Wall Member, W; Wyota Member, Wy; Dubuque Formation, Dub; Sinsinawa Member, S; Stewartville Member, St; Frankville, Fk; Luana, L; and Littleport, Lp. K-bentonite beds indicated by arrow heads on right side of column; I-1 etc. labels refer to K-bentonite beds identified by Mossler and Hayes (1966) in northeastern Iowa.

Figure 12. Carimona Limestone (Cm) overlain by the Decorah Shale (Dc) at the Summit Avenue outcrops in St. Paul, Minnesota (locality 89). Meter rule marks the position of the Deicke K-bentonite Bed (D). Warren D. Huff stands at the level of the Millbrig K-bentonite Bed (M).

Overlying the Castlewood/Carimona Member is the Spechts Ferry Shale Member of the Decorah containing the Millbrig K-bentonite Bed (I-2 bentonite of Mossler and Hayes, 1966). The Millbrig can be traced from its type section near Galena in northwestern Illinois (locality 5) northwestward through the outcrop belt to Minneapolis, Minnesota (Kolata et al., 1986). Average thickness is about 5 cm. The Millbrig continues into the subsurface between the outcrop belts of the Upper Mississippi Valley and eastern Missouri. The bed is present in cores in Jones and Louisa Counties, Iowa, and in Hancock County, Illinois (localities 95, 96, and 112). The Millbrig is not easily distinguished on wireline logs because there is little petrophysical difference between it and the enclosing shale. In northwestern Illinois and southwestern Wisconsin, the Millbrig and Deicke K-bentonites pinch out on the western flank of the Wisconsin Arch (Kolata et al., 1986). Locally, the interval between the two beds is <20 cm thick suggesting that the stratigraphic interval is condensed as a result of slow depositional rates (Fig. 13). Both beds are absent in south-central Wisconsin, north-central Illinois, and northern Indiana apparently due to uplift of the Wisconsin and Kankakee Arches during late Turinian to early Chatfieldian

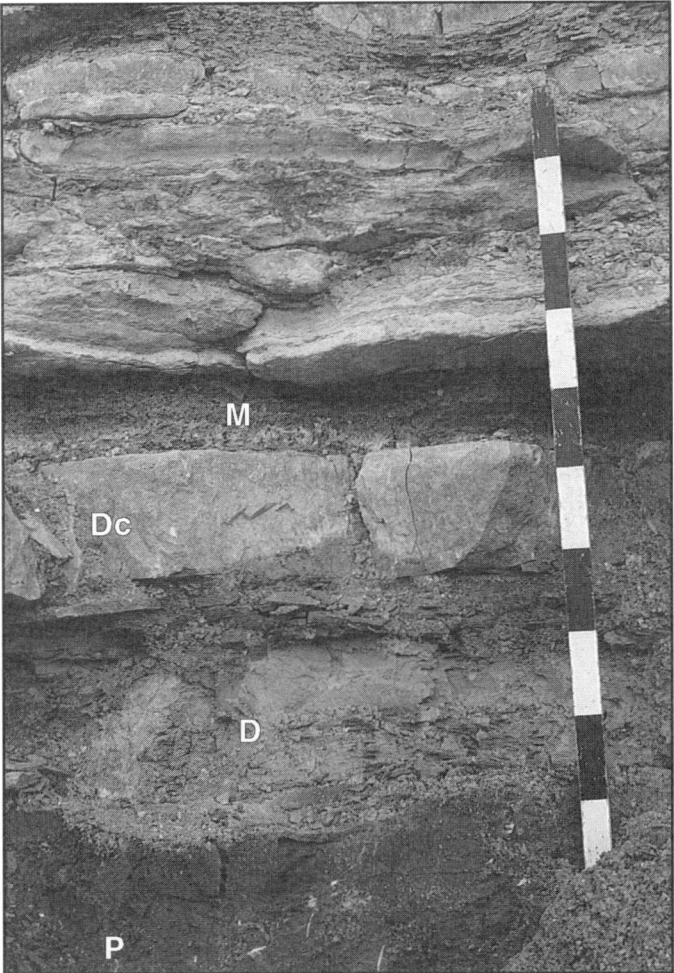

Figure 13. Outcrop near Annaton, Wisconsin (locality 214) of the uppermost limestone bed of the Platteville Group (P) overlain by 25 cm thick Deicke K-bentonite (D), 20 cm of limestone of the Decorah Formation (Dc), and 10-cm-thick bed of the Millbrig K-bentonite (M).

undatus Midcontinent Chronozone. The Elkport K-bentonite Bed (I-3 bentonite of Mossler and Hayes, 1966) lies above the Millbrig commonly 25 cm above the base of the Guttenberg Limestone Member of the Decorah Formation in northwestern Illinois, southwestern Wisconsin, and northeastern Iowa (Fig. 11). The bed generally is less that than 3 cm thick and rarely is present north of Decorah, Iowa, where the Guttenberg grades from limestone to shale. Excellent exposures of the Elkport can be seen near Dickeyville, Wisconsin, and McGregor, Iowa (localities 6, 7, and 12). The Elkport is present in the subsurface of Clayton County, Iowa (locality 116). Like the Deicke and Millbrig, the Elkport pinches out from west to east on the flank of the Wisconsin Arch in southwestern Wisconsin and northwestern Illinois.

The Dickeyville K-bentonite Bed (I-4 bentonite of Mossler and Hayes, 1966) is present locally in the mudstone and wackestone facies of the Guttenberg Limestone Member of the Decorah Formation about 2 m to 3 m above the Elkport K-bentonite. At its type section near Dickeyville, Wisconsin (Fig. 14), the bed is 3 cm thick and lies about 5 m above the Platteville Group. East of there the Spechts Ferry Shale Member oversteps the Platteville gradually pinching out, and from the zero line of the Spechts Ferry eastward the Guttenberg Limestone oversteps the Platteville also decreasing in thickness. In parts of southern Wisconsin, the Guttenberg Dolomite, containing the Dickeyville K-bentonite, rests on the Platteville Group (Fig. 15). Near Polo, Illinois (locality 8), the Dickeyville rests directly on top of the Platteville with a thin interval of Guttenberg lying above it. These stratigraphic relations clearly show the overstepping nature of the Decorah Formation from west to east on the flank of the Wisconsin Arch (Kolata et al., 1986). The Elkport and Dickeyville

time resulting in erosion and/or nondeposition of the volcanic ash. Kay (1931) believed the Hounsfield "metabentonite" of upstate New York to be equivalent to what we now know as the Millbrig (Willman and Kolata, 1978).

The Deicke and Millbrig have been correlated by chemical fingerprinting from southeastern Minnesota to southeastern Missouri (Kolata et al., 1986, 1987) and by wireline logs from Missouri to the southern Appalachian Mountains and into the Michigan Basin and southern Ontario (Huff and Kolata, 1990; Kolata et al., 1990). Both beds also have been correlated on the basis of phenocryst mineralogy from the Cincinnati Arch east and southeast through the Valley and Ridge Province of the Appalachians (Haynes, 1994). Furthermore, there is some suggestion that the Millbrig may be equivalent to the Kinnekulle K-bentonite (Bergström et al., 1995) in Baltoscandia (Huff et al., 1992). The Deicke and Millbrig occur in the *Phragmodus*

Figure 14. Stratigraphic succession consisting of the Platteville Group (P) and Spechts Ferry Shale (SF) and Guttenberg Limestone Members (Gt) of the Decorah Formation exposed near Dickeyville, Wisconsin (locality 6). The Deicke (D) and Millbrig K-bentonite Beds (M) are situated near the base of the Spechts Ferry Shale just above the level of the person's head. The Elkport K-bentonite (E) is at the base of the overhanging Guttenberg Limestone and the Dickeyville K-bentonite (Dk) is situated in the limestone near the top of the outcrop.

Figure 15. Platteville Group (P), Guttenberg Dolomite Member (Gt) of the Decorah Formation and basal dolomite beds of the Dunleith Formation (Dl) of the Galena Group exposed in a roadcut near Dodgeville, Wisconsin (locality 215). Top of meter rule marks the top of the Platteville. The Dickeyville K-bentonite (Dk) is a 5-cm-thick clay bed forming a prominent reentrant within the Guttenberg Dolomite 1.5 m above the Platteville Group.

have been correlated by chemical fingerprinting methods in the Upper Mississippi Valley region (Kolata et al., 1986).

The Dunleith Formation in the Upper Mississippi Valley region contains several K-bentonites, some of which are persistent and widespread (Fig. 11). An unnamed bed 2 cm thick is present locally in southeastern Minnesota (locality 9) and northeastern Iowa (locality 11) near the middle of the Fairplay Limestone Member. The Calmar, Conover, and Nasset K-bentonite Beds, in ascending order, occur at many localities in their type area of northeastern Iowa and southeastern Minnesota (Willman and Kolata, 1978). The Calmar, ranging from 2 cm to 6 cm thick, commonly is present within a 30 cm shaly zone near the middle of the Rivoli Limestone Member. The bed was informally called "K" by Levorson and Gerk (1972) and is well exposed locally in Iowa and Minnesota (localities 10, 13, 14, 15, and 16). The Conover K-bentonite is present locally in northeastern Iowa at the top of the Rivoli Limestone Member (localities 16, 17, and 19). Typically, the bed is about 2 cm thick, but near Volney, Iowa (locality 17), it is 6 cm. The Nasset K-bentonite (I-5 bentonite of Mossler and Hayes, 1966) occurs in the lower half of the Sherwood Limestone Member in outcrops near Decorah and Guttenberg, Iowa (localities 10 and 11), and Wykoff, Minnesota (locality 18), and ranges from 2 cm to 7.5 cm thick. The Calmar, Conover, and Nasset K-bentonites are all exposed in a quarry near Waukon, Iowa (locality 19). In southern Wisconsin and northern Illinois, where these K-bentonites are thin or absent, their apparent position in outcrops is commonly marked by a prominent bedding plane. The Dunleith K-bentonites were correlated with those in the Shoreham-Denmark succession in New York (Templeton and Willman, 1963), but this correlation has not been tested by chemical fingerprinting.

The Haldane K-bentonite Bed (I-6 bentonite of Mossler and Hayes, 1966) is present near the middle of the Wall Limestone Member of the Dunleith Formation over a wide area of the Upper Mississippi Valley region (Willman and Kolata, 1978). The type section is situated in northern Illinois (locality 20), and the bed has been observed as far north as Rochester, Minnesota (locality 21). The thickness of the Haldane is typically 4 cm, but it is as much as 10 cm near Dixon, Illinois (locality 22). An unnamed K-bentonite as much as 5 cm thick has been observed locally near the top of the Wyota Limestone Member in northern Illinois, southern Wisconsin, and southern Minnesota (Templeton and Willman, 1963; Willman and Kolata, 1978). The Dunleith K-bentonites occur in the *Plectodina tenuis* Midcontinent Chronozone (Thomas H. Shaw and Walter C. Sweet, 1994, written communication).

The Dygerts K-bentonite Bed (I-7 bentonite of Mossler and Hayes, 1966), in the lower part of the Sinsinawa Dolomite Member of the Wise Lake Formation, is one of the most widespread and persistent K-bentonites in the Upper Mississippi Valley region. At the type section in Galena, Illinois (locality 23), the 3 cm thick bed is situated 7.3 m above the base of the Wise Lake. It is exposed in quarries near Decorah, Iowa (localities 24 and 25), Rochester, Minnesota (locality 21), and Loves Park, Illinois (locality 26). The Dygerts is believed to be one of two K-bentonite beds recovered from test boring of the Wise Lake Formation in De Kalb, Kane, Kendall, and Du Page Counties in northeastern Illinois (Kempton et al., 1987a, 1987b; Curry et al., 1988). The beds occur at approximately 24 m and 30 m below the top of the Galena Group and although they are only 2 cm to 3 cm thick, they produce a pronounced deflection on many of the gamma-ray/neutron logs. Judging from the stratigraphic relations, the Dygerts is probably the deeper of the two beds. The Wise Lake Formation contains the *Belodina confluens* Midcontinent conodont fauna (Thomas H. Shaw and Walter C. Sweet, 1994, written communication).

Four unnamed K-bentonite beds are present locally in the Dubuque Formation in southeastern Minnesota and northeastern Iowa (Fig. 16). Two of the beds in Fillmore County, Minnesota (locality 27) were described by Weiss (1954) as feldspathized bentonite beds as much as 10 cm thick. In a detailed stratigraphic analysis, Levorson et al. (1979) described the occurrence of four K-bentonites in the Dubuque Formation in Iowa and Minnesota, including the two noted by Weiss (1954). All are present in a quarry near Spring Valley, Minnesota (locality 28). In ascending order these include (1) a 4 cm bed at the top of the Frankville Limestone Member, (2) a 2 cm bed in the middle of the Luana Limestone Member, (3) a 7 cm bed in the upper part of the Luana (I-8 bentonite of Mossler and Hayes, 1966; "lower" bentonite of Weiss, 1954), and (4) a 1 cm bed in the upper part of the Littleport Limestone Member ("upper" bentonite of Weiss, 1954). The uppermost Dubuque K-bentonite is exposed near Kendallville, Iowa (locality 29), but no other Dubuque beds are known to occur in the Upper Mississippi Valley region south of here.

The youngest known Ordovician K-bentonite in the region is in Fillmore County, Minnesota, where a 10 cm thick bed is

Figure 16. Dubuque Formation exposed near Spring Valley, Minnesota (locality 216). Four thin, unnamed K-bentonite beds (white arrows) form prominent reentrants near middle and upper part of the quarry high wall.

Figure 17. Outcrop belt of eastern Missouri (shaded) and Illinois Basin region.

present in the lower part of the Maquoketa Group (localities 30 and 31). The Dubuque and Maquoketa K-bentonites are in the *A. superbus* and *A. ordovicicus* conodont Zones and are among the youngest Ordovician K-bentonites known in North America (Fig. 11). Two K-bentonites were noted by Templeton and Willman (1963) from the Brainard Formation in northern Illinois, but no localities were given, and we were not able to confirm their existence in our search of outcrops and subsurface samples. We also were unable to find K-bentonites in the Ordovician outcrop belt of eastern Wisconsin.

Eastern Missouri

Ordovician K-bentonite beds are exposed in numerous roadcuts and quarries in the area of the Lincoln Fold and on the east flank of the Ozark Dome, all in eastern and southeastern Missouri (Fig. 17). In contrast to the Upper Mississippi Valley, this region has more K-bentonites in the Turinian Plattin Group (=Platteville Group) and fewer in the Chatfieldian Kimmswick Limestone (=Galena Group). The Kimmswick consists predominantly of grainstones deposited in relatively high energy environments. Volcanic ash falling in this region apparently was dispersed by current action. We would expect only the largest ash falls to be preserved.

An Ibexian K-bentonite (Fig. 18) has been reported in the subsurface approximately 60 m below the top of the Powell-Smithville Formation in southeastern Missouri (McCracken, 1955). The bed, believed to occur in the *Oepikodus evae* North Atlantic Conodont Zone (Thomas H. Shaw, 1993, written communication), apparently extends into the subsurface of the Illinois Basin (Howard R. Schwalb, 1985, personal communication).

A late Whiterockian K-bentonite is present locally in the Dutchtown Limestone in southeastern Missouri (Thomas H. Shaw, 1993, personal communication). Sample studies on file at the Missouri Geological Survey report a K-bentonite containing biotite at 34 m and 37 m below the top of the Dutchtown Limestone in two wells in Cape Girardeau County, Missouri (localities 101 and 102). The exact biostratigraphic position has not been determined, but Midcontinent Conodont Faunas 5 and 6 have been reported from the Dutchtown (Repetski, 1973).

Locally, the Mohawkian (Turinian Stage) Plattin Group

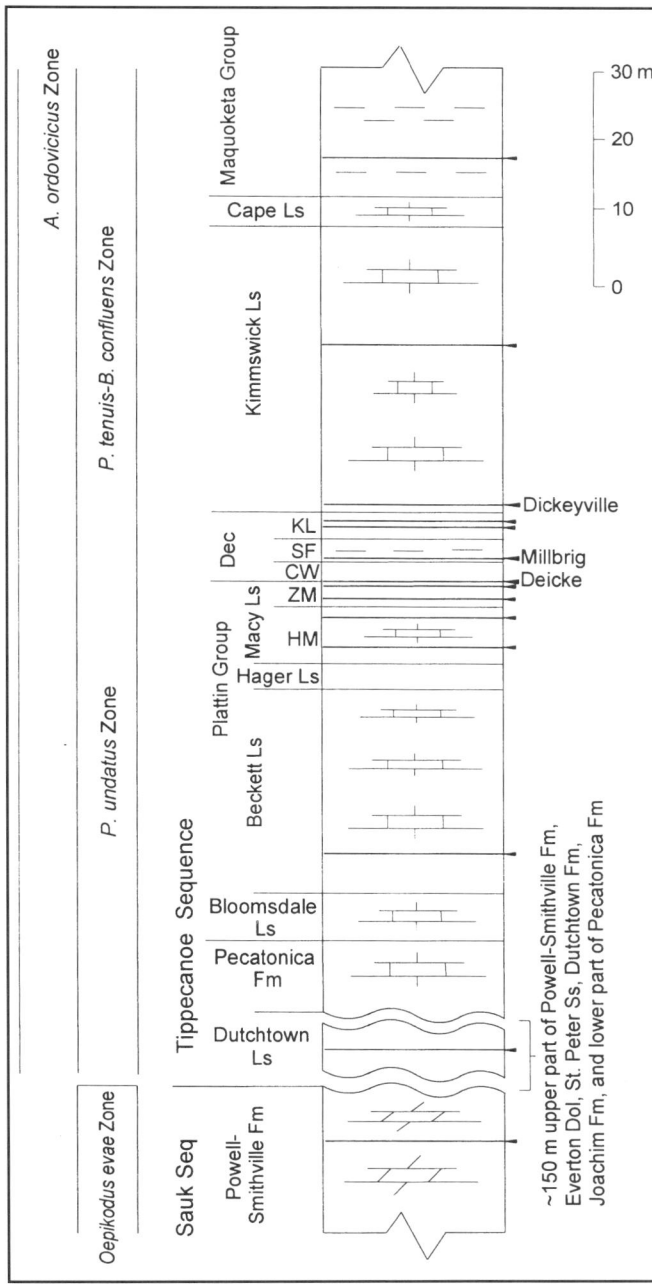

Figure 18. Generalized stratigraphic column for eastern Missouri region. Abbreviations: Hook Member, HM; Zell Member, ZM; Decorah, Dec; Castlewood Member, CW; Spechts Ferry Member, SF; and Kings Lake Member, KL; dolomite, Dol; limestone, Ls; and sandstone, Ss.

(Thompson, 1991) of southeastern Missouri contains at least four K-bentonite beds. A 5 cm thick bed is present in the Beckett Limestone 5.5 m above the shaly strata at the top of the Bloomsdale Limestone (approximately 40 m below the Deicke K-bentonite) in a roadcut along I-55 south of Pevely, Missouri (locality 32). A deflection on the gamma-ray log from the Laclede Gas Company No. 1 Mintert well (depth 1,379 feet) in St. Charles County, Missouri, is in approximately the same stratigraphic position (43 m below the Deicke) and is the petrophysical response to this bed (Plate 2). The Beckett Limestone contains the *P. undatus* Midcontinent conodont fauna.

Three K-bentonites are present in the Macy Limestone in a key outcrop on U.S. Highway 61 west of Ste. Genevieve, Missouri (locality 33). The lowest bed, unusually thick at 15 cm, is near the middle of the Hook Limestone Member, 15.5 m below the top of the Plattin Group. A second K-bentonite, as much as 5 cm thick, is situated 9.7 m below the top of Plattin. This same bed is believed to be exposed in the Hook Limestone in an I-55 roadcut immediately north of Pevely, Missouri (locality 34), where it lies 9 m below the Deicke K-bentonite. A third bed, exposed at the Highway 61 roadcut, occurs in the Zell Limestone Member at 5 m below the top of the Plattin. The *P. undatus* Chronozone is believed to extend through the upper half of the Macy Limestone, including the Zell Limestone Member, to near the top of the overlying Decorah Formation (Thomas H. Shaw and Walter C. Sweet, 1994, written communication).

The Deicke and Millbrig are present in the Castlewood Limestone Member and Spechts Ferry Shale Member of the Decorah, respectively, at many outcrops in eastern Missouri. The type section for the Deicke is situated in the bluffs of the Meramec River west of St. Louis (locality 4). An excellent reference section for both the Deicke and Millbrig is a roadcut on the north frontage road of I-44, west of the Allenton–Six Flags amusement park exit (locality 35). Other well-exposed sections are described by Kolata et al. (1986) including the Eureka section shown in Figure 19. Both the Deicke and Millbrig are thicker in eastern Missouri than in the Upper Mississippi Valley outcrop belt. Furthermore, they show a regional southeastward thickening from north of St. Louis, where they generally are 5 cm to 10 cm (locality 38), to near Ste. Genevieve (locality 39) where the Millbrig is 15 cm and at Illmo (locality 40) where the Deicke is 45 cm. A notable section near Bloomsdale (locality 37) exposes a 3 cm K-bentonite 0.6 m below the Deicke that has not been observed elsewhere in the region. The Deicke and Millbrig typically occur as light to medium gray, soft to compact clay (Fig. 20), but locally the beds are feldspathized and may contain abundant molds and casts of macrofossils (Fig. 21).

The Deicke can be traced with confidence on wireline logs from near its type section west of St. Louis, Missouri, eastward through the Illinois and Michigan Basins into the Appalachian Basin. In the Laclede Gas Company No. 1 Mintert well in St. Charles County (locality 105), the Deicke is marked by a prominent deflection on the gamma-ray log about 1.7 m below the base of the Decorah Formation (Plate 2). The Millbrig, widespread in the eastern Missouri outcrop belt, is not readily identified on wireline logs here, because it is obscured by the enclosing shaly strata at the base of the Spechts Ferry Shale Member. The interval of limestone between the Deicke and the Spechts Ferry Shale gradually thickens southward through Monroe County, Illinois, and Perry and Cape Girardeau Counties, Missouri. Thickening of the interval corresponds to the overall

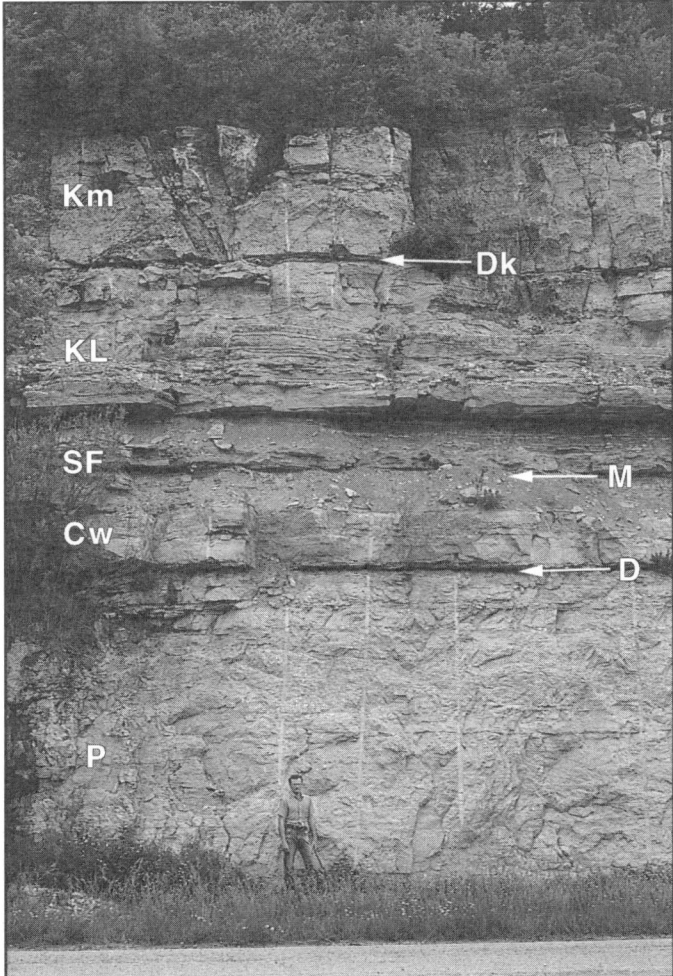

Figure 20. Exposure of the Millbrig K-bentonite (M) near House Springs, Missouri (locality 218). Texture and color vary from soft, plastic, light gray clay on the left through hard, flaky, medium gray in the middle to brittle, medium gray on the far right. Variations appear to be the result of alteration by ground-water.

Figure 19. Exposure of the Platteville Group (P), Decorah Formation, and Kimmswick Limestone (Km) (in ascending order) near Eureka, Missouri (locality 217). The Deicke K-bentonite Bed (D) forms a prominent reentrant near the top of the massively bedded Platteville limestones and is overlain by the Castlewood Limestone (Cw), Spechts Ferry Shale (SF), and Kings Lake (KL) Members of the Decorah Formation. The Millbrig K-bentonite (M) forms a prominent reentrant near the middle of the Spechts Ferry Shale Member (SF) of the Decorah Formation and the Dickeyville K-bentonite (Dk) forms a reentrant in the Kimmswick Limestone 2 m to 3 m below the top of the outcrop. Daniel B. Blake stands at base of outcrop.

thickening of the Plattin Group toward its depocenter situated in the southern Illinois Basin (Kolata and Noger, 1991). At Gray's Point quarry (locality 40) in Scott County, the southernmost exposure of the upper Plattin Group in Missouri, the Deicke is 45 cm thick and occurs 8 m below the Decorah Formation. The Millbrig is absent at this locality. The overall stratigraphy in Gray's Point quarry is quite similar to that in the Shell Oil Co. No. 1 Trail of Tears State Park well (locality 148) in nearby Cape Girardeau County (Plate 5).

At least two K-bentonites, each less than 5 cm thick, have been observed locally in the Kings Lake Limestone Member, Decorah Formation (Kolata et al., 1986). These are exposed at the Bloomsdale outcrop (locality 37).

Our subsurface studies suggest that the House Springs K-bentonite Bed (Kolata et al., 1986) may be equivalent to the Dickeyville K-bentonite, in the Guttenberg Limestone Member of the Decorah Formation in southwestern Wisconsin (Willman and Kolata, 1978). In this book, we correlate the Dickeyville with the House Springs K-bentonite of eastern Missouri and with a thick and widespread K-bentonite in the southeastern United States. However, this correlation must be considered provisional until more definitive subsurface information is available, particularly in the critical area between Missouri and Wisconsin. The Dickeyville occurs near the base of the Kimmswick Limestone, commonly about 1 m above the Guttenberg Limestone (Fig. 19). Throughout most of the region, it ranges in thickness from 0.5 cm to 3 cm but is as much as 15 cm thick near Altenburg, Perry County (locality 36). Where the bed has no measurable thickness, its position is commonly marked by a prominent bedding plane. The Dickeyville is present in two Ste. Genevieve County, Missouri, cores (localities 103 and 104). The bed lies in the upper part of the *P. undatus* Chronozone.

Another Kimmswick K-bentonite, 6 cm thick and occurring 20 m below the base of the Maquoketa Group, has been observed at a single locality near Clarksville (locality 41). In a nearby outcrop along Calumet Creek, an unnamed K-bentonite is present in the Richmondian Maquoketa Group (locality 42).

Illinois Basin

The Illinois Basin covers an area of approximately 285,000 km² in parts of Illinois, Indiana, and Kentucky and contains a thick succession of Ordovician rocks confined entirely to

Figure 21. Feldspathized Millbrig K-bentonite Bed from the Decorah Formation at Warrenton, Missouri (locality 219), containing abundant molds and casts of the orthid brachiopod *Pionodema subaequata*. Magnification ×1.3.

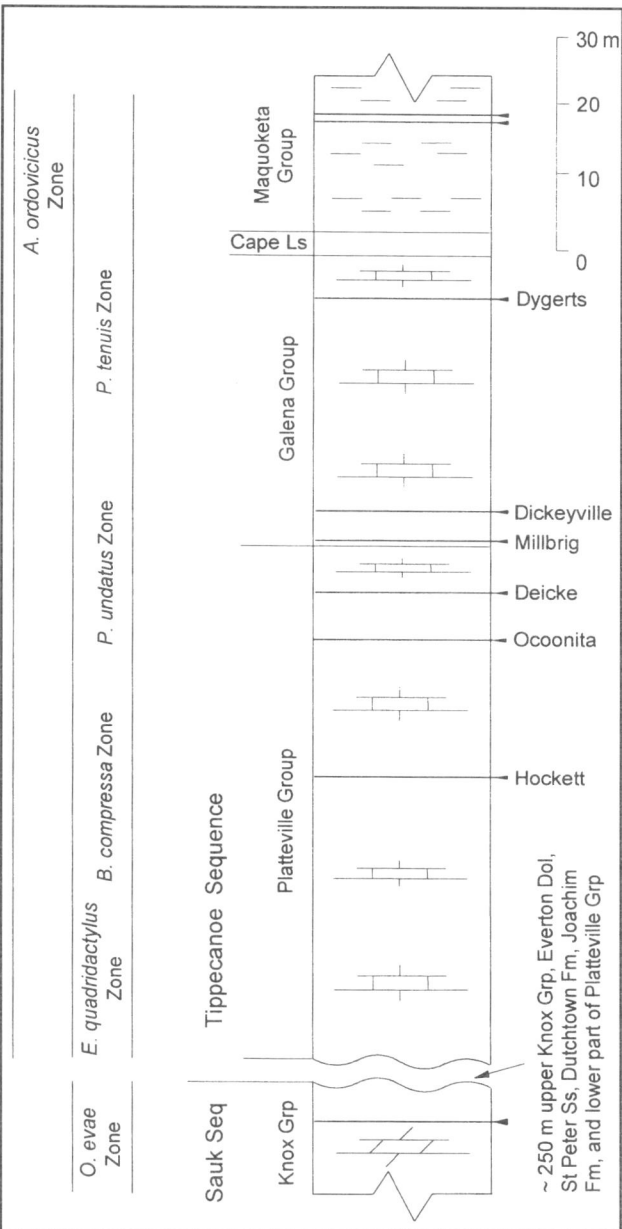

Figure 22. Generalized stratigraphic column for the Illinois Basin. K-bentonites indicated by arrow heads on right side of column. Abbreviations: dolomite, Dol; limestone, Ls; and sandstone, Ss.

the subsurface (Figs. 17 and 22). At the basin depocenter in southern Illinois, the top of the Ordovician succession occurs at depths of as much as 2,000 m below surface. Researchers have observed K-bentonites in cores, drill cuttings, and on wireline logs from this area.

The oldest known K-bentonite is identified from small chips and flakes of clay in drill cuttings from the upper part of the Ibexian Shakopee Dolomite of the Knox Group in the subsurface of Franklin, Pope, and Union Counties (Fig. 22) in southern Illinois (Howard R. Schwalb, 1985, personal communication). In the Humble Oil and Refining Co. No. 1 J. F. Pickel and the Texas Pacific Oil Co. No. 1 Mary L. Streich wells, the position of the K-bentonite bed is marked by an abrupt increase in gamma-ray counts, high porosity, and relatively long acoustic traveltimes (see notes under localities 97, 98, and 100). The bed is approximately in the same stratigraphic position as a K-bentonite described by McCracken (1955) in the Powell-Smithville Formation in southeastern Missouri and may be equivalent to it.

Wireline logs from wells that penetrate the pure carbonate rocks of the Platteville Group (=Black River Group) in the Illinois Basin show several prominent and persistent deflections that appear to correspond to K-bentonite beds (Plate 3). The oldest bed is marked by a deflection on the gamma ray log in Perry County, Illinois (locality 120), at 4,878 ft and Washington

County (locality 121), Illinois, at 5,090 ft. Higher in the Platteville in Montgomery County (locality 124), two apparent K-bentonites are marked by prominent deflections at 2,836 ft and 2,860 ft, one of which is present also in the Marion County well (locality 122) at 4,718 ft.

A widely occurring deflection on gamma-ray logs appears to mark a K-bentonite bed at a consistent depth of between 10.7 m and 15.2 m below the top of the Platteville (Plate 3). This bed is particularly well developed in Marion, Clinton, Douglas, Piatt, and Champaign Counties, Illinois (localities 122, 123, 125, 127, and 128). The bed extends through the eastern part of the Illinois Basin in Wayne and Lawrence Counties, Illinois, through Vigo and Decatur Counties, Indiana (localities 176 and 180), into southwestern Ohio where it is equivalent to K-bentonite bed γ of Stith (1979). Likewise, the bed can be traced on wireline logs through Oldham County, Kentucky, and Switzerland County, Indiana, into southwestern Ohio (localities 154 and 155).

The Deicke K-bentonite is situated near the top of the Platteville Group in southern Illinois, western Kentucky, and southern Indiana. It is marked by a persistent deflection on gamma-ray/neutron logs 4 m to 8 m below the overlying Galena Group (=Trenton Group). From its type section near St. Louis, Missouri, the Deicke can be correlated with a high degree of confidence through the subsurface of the Illinois Basin. The bed was encountered in the cored Superior Ford No. C-17 H. C. Ford et al. well in White County, Illinois (locality 119), at 8 m below the top of the Platteville. A gamma-ray log is not available for this drill hole, but the stratigraphy in the core corresponds closely to that observed on wireline logs from the nearby Texaco Inc. No. 1 J. M. Walters (Plate 5) and Conoco No. 1 Einar Dyhrkopp wells in Gallatin County, Illinois (localities 150 and 99). Whole-rock analysis of the K-bentonite in the Superior Ford well show it to be chemically indistinguishable from the Deicke.

In the northwestern part of the basin, the Deicke is thin or absent, with little or no expression on wireline logs. North of LaSalle County, Illinois, the bed pinches out on the southern flanks of the Wisconsin Arch. Likewise, in northeastern Illinois and northern Indiana the Deicke is thin or absent apparently as the result of erosion or nondeposition on the crest of the Kankakee Arch during late Turinian time. The stratigraphic relations are shown in cross section (Plate 3). Northeast of Iroquois County, Illinois (locality 129), the interval between the Deicke and the top of the Platteville diminishes to less than one meter in Jasper County, Indiana. In Lake County, Indiana (locality 131), the Deicke is absent and the overlying, slightly argillaceous carbonate rocks of the Galena Group (=Trenton) rest on Platteville strata that are as old or older than the Deicke K-bentonite Bed. The Deicke is present on the northern flank of the Kankakee Arch in the Michigan Basin approximately 70 km northeast of Lake County, Indiana, in the Security Oil and Gas Co. No. 1 Thalmann well in Berrien County, Michigan (locality 132). The bed can be traced eastward through the subsurface in the southern part of the Michigan Basin where it is referred to as the "Black River shale" (Lilienthal, 1978).

Within the southern Illinois Basin, the interval of limestone between the Deicke and the base of the Galena (Trenton) thickens gradually into the center of the basin from 1.7 m in St. Charles County, Missouri, to 7.6 m in Gallatin County, Illinois, and back down to 4.5 m at the eastern margins of the basin in Switzerland County, Indiana (Plate 5). These thickness trends show that carbonate sedimentation kept pace with relatively rapid subsidence of the basin during late Turinian time.

The Millbrig K-bentonite Bed in the outcrop belt in southeastern Missouri consistently lies in the basal shaly strata of the Decorah Formation. On wireline logs it is not readily distinguishable from the enclosing shale. North and east of the Missouri outcrop belt, the Decorah grades abruptly to argillaceous limestone and the Millbrig appears on wireline logs as a single prominent deflection at or near the contact between the Platteville and Galena Groups. The Millbrig is particularly well developed in Champaign, Piatt, Dewitt, and Marion Counties, Illinois (Plate 3). It is present in the eastern part of the Illinois Basin in Vigo, Clay, Lawrence, and Decatur Counties, Indiana, and continues into southwestern Ohio where it correlates with marker bed α of Stith (1979). The Millbrig is equivalent to the "Mud Cave bentonite" in Kentucky and the T-4 in Tennessee.

The Dickeyville K-bentonite Bed is traceable on wireline logs from southeastern Missouri northward and eastward through the Illinois Basin (Plates 3, 4, and 5). It produces a moderate deflection on gamma-ray/neutron logs on the western side of the basin in Pike, Brown, and Schuyler Counties (localities 107, 108, 109, and 110), in the northern part of the basin in Douglas, DeWitt, Piatt, Champaign, and Iroquois Counties (localities 125, 126, 127, 128, and 129), and across the southern part of the basin in Franklin, Hamilton, Wayne, and Lawrence Counties, Illinois, and Clay and Greene Counties, Indiana (localities 97, 173, 174, 175, 177, and 178). In the Illinois Basin, the Dickeyville generally occurs between 4.5 m and 7.5 m above the top of the Platteville (=Black River) Group.

The succession of K-bentonites including the Deicke, Millbrig, and Dickeyville show little or no response on wireline logs along the northeastern margin of the Illinois Basin in the northern counties of Indiana (north of a line from Jay to Jasper Counties) and east into northwestern Ohio. This region lies at the juncture of the Kankakee and Cincinnati Arches. Absence of these beds appears to be associated with structural movement on the arches resulting in nondeposition or erosion of the volcanic ash beds. The K-bentonites are present, however, a short distance north of the Indiana and Ohio state lines in the Michigan Basin.

Several K-bentonite beds are present in the upper half of the Galena Group in the subsurface of the Illinois Basin. One of these is marked by a gamma-ray deflection 17 m below the top of the Galena in the subsurface of Pike County, Illinois (Plate 2; localities 107 and 108). This very likely is the same bed exposed nearby in a quarry at Clarksville, Missouri (locality 41), where it is 6 cm thick and situated 20 m below the top of the Galena. In the subsurface of Washington County, Iowa (locality 115), two likely K-bentonite beds are situated at 21 m and 27 m below the top of the Galena, one of which is probably equivalent to the bed observed at Clarksville.

Gamma-ray logs from wells in Williamson, Wayne, and Lawrence Counties, Illinois (localities 172, 174, and 175), show a probable K-bentonite between 3 m to 9 m below the top of the Galena (Plate 4). The relative distance of this bed above the base of the Galena is approximately the same as that observed in the Clarksville K-bentonite described above, and it may be the same bed.

Two K-bentonite beds were reported from the Late Ordovician Scales Formation of the Maquoketa Group near Kentland, Indiana (locality 43), at the northern margin of the Illinois Basin (Templeton and Willman, 1963). A review of the Templeton and Willman field notes in the files of the Illinois State Geological Survey indicates that these beds are 2 cm to 4 cm thick and are separated by 0.6 m of shale at approximately 17 m above the base of the Maquoketa. No K-bentonite beds were observed during an examination of the Kentland Quarry in the spring of 1994.

Central and northern Kentucky and southern Ohio

Ordovician K-bentonite–bearing strata are present in the subsurface of central and northern Kentucky and southern Ohio and in outcrops on the crest of the Jessamine Dome in north-central Kentucky (Fig. 23). Wireline logs recording the presence of ash beds are available from numerous drill holes in the region. Furthermore, a significant amount of information has been published on the occurrence of K-bentonites in this region (Cressman, 1973; Cressman and Noger, 1976; Huff, 1983; Conkin and Dasari, 1986; Conkin and Conkin, 1983, 1992; Stith, 1979, 1986; Huff and Kolata, 1989; Schumacher and Carlton, 1991; Young, 1940; Wickstrom et al., 1992).

A composite stratigraphic column for north-central Kentucky and southern Ohio is shown in Figure 24. Two K-bentonites are present in the Ibexian Beekmantown Dolomite, Knox Group, in the subsurface at two localities in northern Kentucky (localities 203 and 204). They occur at 17 m and 43 m below the top of the Knox Group in Owen County and 22 m and 35 m below the top of the Knox in Gallatin County. They appear to be in the same stratigraphic interval as the Ibexian K-bentonite observed in eastern Missouri and southern Illinois, and one of these could be the same. Furthermore, the two Beekmantown K-bentonites may be equivalent to two K-bentonite beds that have been reported in drill cuttings recovered from the upper part of the Knox Group in a drill hole in the Black Warrior Basin of northern Mississippi (Thomas, 1972, 1988). Conodonts of the *Oepikodus communis* Zone are associated with the K-bentonites (Alberstadt and Repetski, 1989) indicating that they are equivalent in age to the Ibexian K-bentonite in eastern Missouri and the Illinois Basin and probably to those in the Beekmantown.

A K-bentonite was described in drill cuttings from the basal 1.5 m of the late Whiterockian Wells Creek Dolomite (1,752 ft to 1,757 ft) in a sample study of the Jarvis and Marcell No. 1 Parrigan well (locality 205) in Clinton County, Kentucky (Freeman, 1953). The bed lies just above the sub-Tippecanoe unconformity between the Wells Creek and the underlying Knox Dolomite.

Eighteen K-bentonites were identified by Conkin and Conkin (1983) in a detailed study of the Tyrone Limestone, Black River Group, at Boonesborough, Clark County, Kentucky. All occur within the *B. compressa* and *P. undatus* Zones. Some beds are only a few millimeters thick and are a challenge to locate and collect. We suspect that such thin, discontinuous K-bentonites are widespread in Middle Ordovician rocks of eastern North America but cannot be correlated over distances that would allow them to be useful as event-stratigraphic markers. Given the extensive Ordovician volcanism that is known to have occurred along the eastern margin of North America, it is not unreasonable to expect small amounts of K-bentonite to occur on bedding planes throughout the Middle Ordovician succession. For the purposes of this book, the very thin discontinuous K-bentonites are noted but not emphasized in regional correlations.

Five relatively thick and persistent K-bentonites in the upper part of the Tyrone Limestone have been correlated through the region on the basis of chemical fingerprinting (Huff, 1983). These are designated by Stith (1979, 1986), in ascending order, as "b," "a," "γ," "β," and "α" (Plate 6). Their position is marked

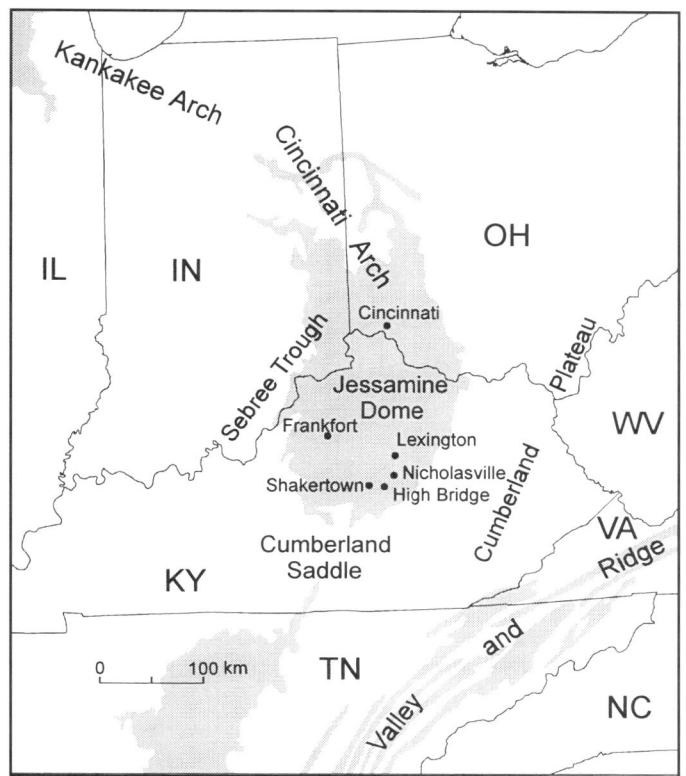

Figure 23. Outcrop belt (shaded) and major structural features in northern Kentucky, eastern Indiana, and southern Ohio.

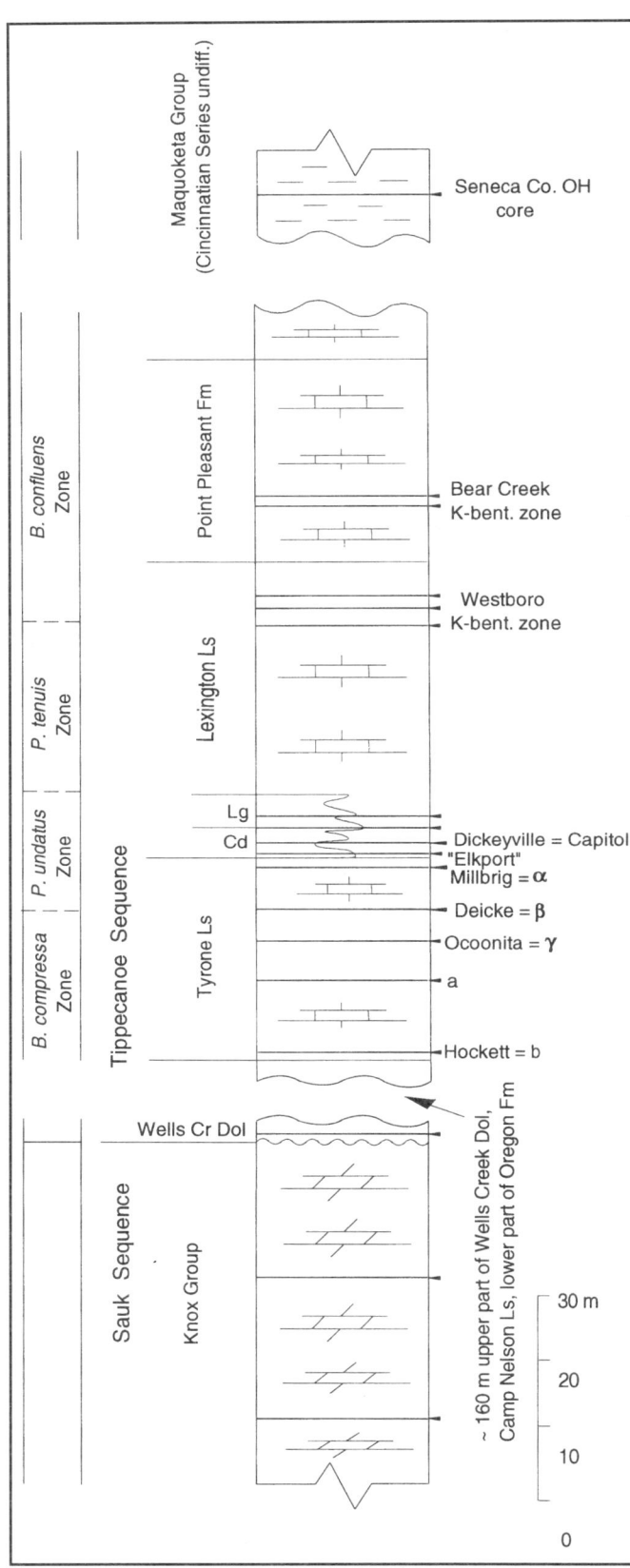

Figure 24. Generalized stratigraphic column for the region of north-central Kentucky and southern Ohio. K-bentonites b, a, γ, β, and α from Stith (1979, 1986). Bear Creek and Westboro K-bentonite zones from Schumacker and Carlton (1991). Abbreviations: Curdsville, Cd; Logana, Lg; dolomite, Dol; and limestone, Ls.

TABLE 1. NOMENCLATURE FOR THE DEICKE AND MILLBRIG K-BENTONITE BEDS		
	Deicke	**Millbrig**
Stith (1979) Ohio	β	α
McFarlan (1943) Kentucky and Tennessee	Pencil cave	Mud cave
Wilson (1949) central Tennessee	T-3	T-4
Fox and Grant (1944) eastern Tennessee	B-3	B-6
Miller and Fuller (1954) southwestern Virginia (western outcrop belt)	R-7	R-10
Huffman (1945)	V-4	V-7
Rosenkrans (1936) southwestern Virginia (central outcrop belt)	V-3	V-4
Bell (1954) southeastern Minnesota	bentonite A	bentonite B
Weiss and Bell (1956) southeastern Minnesota	Carimona bentonite	Spechts Ferry bentonite
Mossler and Hayes (1966) northeastern Iowa	I-1	I-2
Huff (1983) northern Kentucky	3	2
Kay (1931) eastern North America		Hounsfield
Lilienthal (1978) Michigan Basin	Black River shale	

on the gamma log of the Ohio Department of Natural Resources ODS-2626 from Highland County, Ohio (locality 183). The lower three beds typically consist of K-bentonite mixed or interbedded with shale and argillaceous limestone and are approximately in the same stratigraphic position and possibly equivalent to those in the Macy Limestone of southeastern Missouri. The upper two, informally called Pencil Cave and Mud Cave (McFarlan, 1943; see Table 1), are equivalent to the Deicke and Millbrig K-bentonite Beds, respectively (Huff and Kolata, 1990). These are the thickest and most widespread beds in the region and are easily recognized in outcrop, drill cores, and on wireline logs.

The two beds can be distinguished from each other by the kind and proportion of volcanogenic phenocrysts. Common phenocrysts in the Deicke include labradorite and various Fe-Ti minerals and in the Millbrig andesine, quartz, and biotite (Haynes, 1994). Typically, the Millbrig is yellowish gray and has abundant biotite and quartz, whereas the Deicke is greenish gray and lacks obvious biotite and quartz (Haynes, 1992).

The Deicke is variable in thickness over relatively short distances. It ranges from 15 cm at the entrance shaft to the Black River Mine near Carntown, Pendleton County, Kentucky (locality 45), and 25 cm near Frankfort, Kentucky (Fig. 25; locality 220), to as much as 92 cm near Nicholasville, Kentucky (Fig. 26; locality 44). Based on a detailed study of three cores from Butler and Clermont Counties in southwestern Ohio, Stith (1986) suggested that the Deicke (marker bed β) consists of two distinct K-bentonite layers separated by 0.4 m to 0.8 m of micrite. This could represent the Deicke plus a locally occurring bed within the same interval or, as suggested by Stith (1986), argillaceous bentonitic intervals may represent bottom mixing of perhaps a single thick bed resulting in thin K-bentonite beds separated by clay-rich limestone layers. A similar couplet of K-bentonites is present near Bloomsdale, Missouri (locality 37), where a thin K-bentonite bed occurs 0.6 m below the Deicke. The latter is probably equivalent to the Dead Horse Road Quarry "metabentonite" described by Conkin and Conkin (1992).

The Millbrig shows a patchy distribution in the region of the Jessamine Dome in north-central Kentucky. It is 83 cm thick near Shakertown, Kentucky (Fig. 27; locality 47), 38 cm only 3 km to the northeast at High Bridge (locality 46), and completely absent near Nicholasville (Haynes, 1994). In this region, the Millbrig lies near the contact between the Tyrone Limestone and overlying Lexington Limestone, formations that are separated by a disconformity. Locally, in central Kentucky, pre-Lexington erosion apparently removed the Millbrig (Cressman, 1973). In the subsurface of southwestern Ohio and northern Kentucky, however, there are as much as 2 m of micritic limestone (typical of the Tyrone Limestone) between the Millbrig and the overlying Lexington Limestone (Stith, 1986). Local absence of the Millbrig in this region suggests that the ash either was not deposited or more likely was dispersed by bottom currents and mixed with carbonate sediments.

A wireline cross section from Mason County, Kentucky, to Delaware County, Ohio, shows the K-bentonites in the upper part of the Tyrone, including the Deicke and Millbrig, to be marked by persistent and readily traceable deflections on gamma-ray logs (Stith, 1986). In regions of Ohio that are north of Delaware County, however, it becomes difficult to trace individual K-bentonite beds with confidence (David A. Stith, 1994, personal communication). Local facies changes in the upper part of the Black River Limestone and absence of key K-bentonites in some sections creates uncertainty in bed-for-bed correlations (Wickstrom et al., 1992). For example, in the extensively studied Seneca County drill hole in northwestern Ohio (Wickstrom et al., 1985),

Figure 25. Exposure of the Tyrone Limestone near Frankfort, Kentucky (locality 220). Pick is resting near the base of the Deicke K-bentonite Bed (D).

Figure 26. Entrance to underground quarry in the Tyrone Limestone near Nicholasville, Kentucky (locality 44). The Deicke K-bentonite is 90 cm thick and forms a prominent reentrant near the middle of the high wall above the two adits.

Figure 27. Biotite-rich Millbrig K-bentonite (M) exposed near Shakertown, Kentucky (locality 47). The bed is approximately 70 cm thick, overlies the Tyrone Limestone and is overlain by the Lexington Limestone.

several K-bentonites occur in the upper part of the Black River, but it is unclear how they correlate with those in the subsurface of southern Ohio (David A. Stith, 1994, personal communication). Likewise, northeast of Hocking County, Ohio, abrupt changes in thickness and facies within the succession of strata near the Black River/Trenton contact make it difficult to trace K-bentonites confidently (Ronald Riley, Ohio Division of Geological Survey, 1995, written communication). Consequently, it is unclear how the K-bentonites of southern Ohio correlate northeastward through Ohio into southern Ontario and western Pennsylvania. Ryder (1991, 1992a, 1992b) and Ryder et al. (1992) have attempted a bed-for-bed correlation of certain K-bentonites from northern and central Ohio to central Pennsylvania and eastern West Virginia, but we disagree with Ryder's regional correlations of the Deicke and Millbrig.

K-bentonites are known from the overlying Lexington Limestone (Trenton) and Point Pleasant Formation in northern Kentucky and southwestern Ohio. Twenty-seven K-bentonites, most of which are only 2 mm or 3 mm thick and discontinuous, were reported by Conkin and Dasari (1986) from outcrops of the Curdsville Limestone Member of the Lexington from near Shakertown, Mercer County, Kentucky. A particularly prominent bed, named the Capitol "Metabentonite" by Conkin and Dasari (1986) for outcrops near the Kentucky State Capitol Building at Frankfort, occurs near the middle of the Curdsville and typically consists of 80 cm to 100 cm of intercalated K-bentonite, clayey shales, and thin layers of limestone. This bed has been observed in the subsurface in cores and wireline logs in southern Ohio and northern Kentucky where it typically is present in the lower part of the Trenton Group approximately 12 m above the Deicke K-bentonite.

As many as 12 thin K-bentonites have been reported from the Logana Limestone Member of the Lexington Limestone in Woodford County, Kentucky (Conkin and Conkin, 1992). These authors also report the occurrence of three K-bentonite beds in outcrops of the Grier and Tanglewood Limestone Members of the Lexington in north-central Kentucky.

Cores from northern Kentucky and southwestern Ohio reveal at least three K-bentonite beds near the base of the Lexington Limestone and four near the top named the Westboro K-bentonite zone (Schumacher and Carlton, 1991). The overlying Point Pleasant Formation contains two or three beds referred to as the Bear Creek K-bentonite zone. Those at the top of the Lexington and in the Point Pleasant are represented by thin alternating beds of limestone and K-bentonite together ranging in thickness from 15 cm to 46 cm. The beds occur in the *P. undatus* through basal *B. confluens* Chronozones.

A clay bed was described by Caster and Kjellesvig-Waering (1964) from the upper part of the Cincinnati Group (Richmondian Series) in Adams County, Ohio, but our XRD analyses did not confirm this to be a K-bentonite. However, a confirmed K-bentonite has been found in the Maquoketa Group (undifferentiated Cincinnatian Series) in the Ohio Division of Geological Survey core 2580 from Seneca County, Ohio (Bergström and Mitchell, 1992). This bed occurs at a drilling depth of 915 ft (279 m) in calcareous shale and limestone that lie above a section of Utica Shale containing *G. pygmaeus*.

Central Tennessee

Ordovician rocks are exposed along the crest of the Nashville Dome in central Tennessee (Fig. 28) and are penetrated in the subsurface by numerous exploratory drill holes. The presence of K-bentonite beds in central Tennessee was noted by Wilson (1949), Templeton and Willman (1963), Fetzer (1973), Huff and Kolata (1990), and Haynes (1992, 1994). Several of the thick K-bentonites in north-central Kentucky are traceable on wireline logs southward through the subsurface into central Tennessee (Huff and Kolata, 1990). A composite stratigraphic column of the central and southeastern Tennessee region is shown in Figure 29.

The Rockvale "Metabentonite," occurring in the upper part of the Lebanon Limestone (Conkin and Conkin, 1992) approximately 24 m below the Deicke K-bentonite, is one of the oldest K-bentonites known in central Tennessee (Fig. 29). The bed ranges from 3 cm to 20 cm thick and is exposed at outcrops in DeKalb, Rutherford, and Wilson Counties. The Rockdale does not produce a conspicuous deflection on wireline logs in the central Tennessee region. The bed appears to be in the *B. compressa* Chronozone.

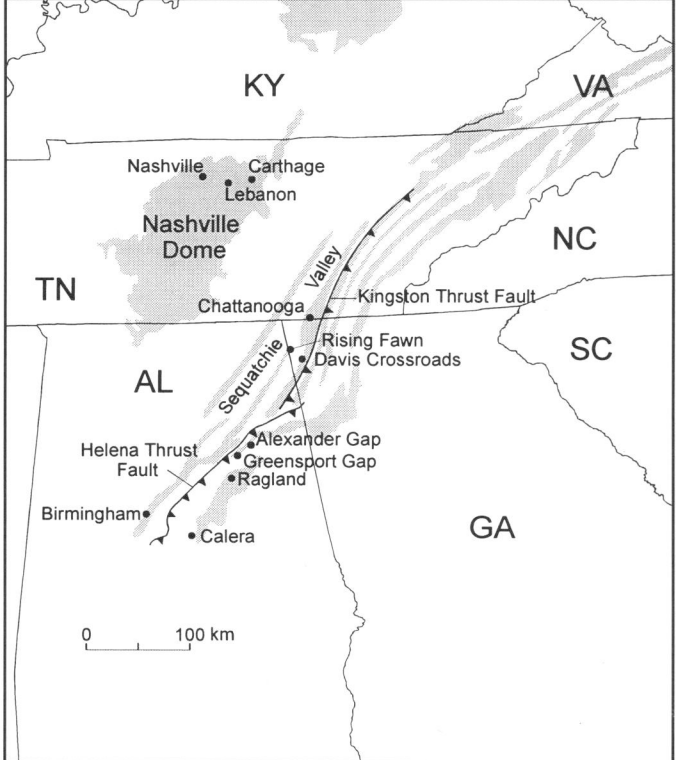

Figure 28. Outcrop belts (shaded) and major structural features in the region of the Nashville Dome and the southern Appalachians.

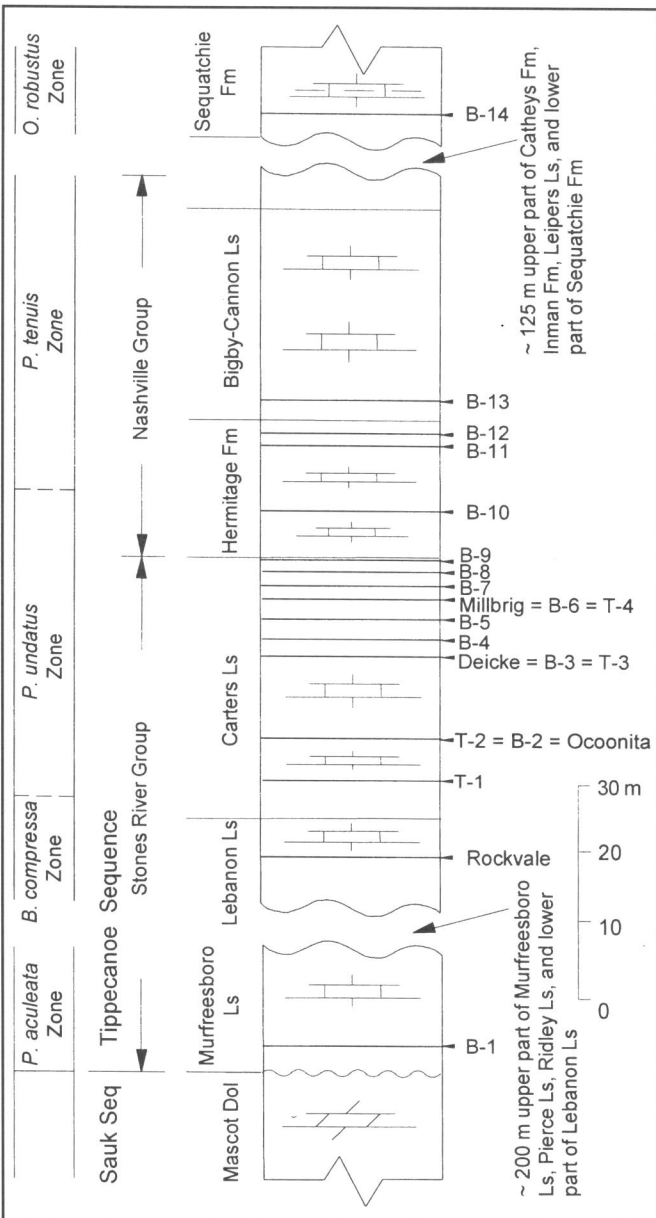

Figure 29. Generalized stratigraphic column for western thrust belts in central and southernmost Appalachians. K-bentonite symbols T-1 through T-4 from Wilson (1949), B-1 through B-14 from Fox and Grant (1944). Abbreviations: sequence, Seq; limestone, Ls; and Formation, Fm.

The Carters Limestone, approximately equivalent to the Tyrone Limestone in Kentucky, contains at least four persistent K-bentonites designated T-1 through T-4 by Wilson (1949). The K-bentonites T-1 and T-2 are locally present in the lower part of the Carters Limestone, primarily in Davidson and Wilson Counties. According to Wilson (1949), T-1 ranges to as much as 20 cm but typically is <5 cm thick and occurs approximately 11 m below the prominent and thick T-3 (Deicke) K-bentonite.

The T-2 ranges from 7 cm to 30 cm thick and consistently lies about 6 m below the T-3.

The T-3 and T-4 are equivalent to the Deicke and Millbrig K-bentonite Beds, respectively (Huff and Kolata, 1990). Typically, the Deicke is 30 cm to 40 cm thick and occurs at the boundary between the upper and lower members of the Carters Limestone over most of the region. The bed is particularly well exposed near South Carthage, Smith County (Fig. 30; locality 48), where it is 40 cm thick. In outcrop, the Deicke has a yellowish green to greenish gray color that distinguishes it from other K-bentonites in the Carters Limestone. A 5 cm to 10 cm thick layer of dark gray to black chert commonly underlies the bed. The Deicke is widespread in the subsurface of central Tennessee, and produces a prominent and widely traceable deflection on wireline logs (Plate 7). The bed is locally absent in Giles, Maury, and Marshall Counties, Tennessee, where the upper Carters Limestone, including the Deicke, was removed by pre-Hermitage erosion (Wilson, 1949). An analysis of core

Figure 30. Roadcut exposing the Deicke K-bentonite at the base of the upper Carters Limestone near Carthage, Tennessee (locality 48). The Millbrig K-bentonite is present at top of the Carters Limestone in the grass covered slope. Thomas E. Guensburg stands at base of outcrop.

and wireline log data from the subsurface of western Tennessee show the Deicke to be thin or absent. In contrast, the bed thickens eastward toward the Valley and Ridge Province.

The Millbrig ranges from 0 cm to 80 cm thick (locality 49) and lies near the contact between the Carters and overlying Hermitage Formation (Fig. 31). Approximately 6 m to 8 m of limestone separate the Millbrig from the Deicke K-bentonite. In some regions of the Nashville Dome, just as it occurs on the Jessamine Dome in north-central Kentucky, the Millbrig is thin or absent as a result of erosion during or prior to deposition of the Hermitage Formation. In outcrop, the Millbrig is buff to light gray in color, contains abundant biotite, and locally consists of two or three graded beds. Where present, the Millbrig produces a prominent deflection on wireline logs. Both the Deicke and Millbrig form a readily identifiable couplet of deflections, particularly on gamma ray/neutron logs in the subsurface of the eastern half of Tennessee.

The stratigraphic position of the Millbrig K-bentonite with respect to the Carters Limestone and Hermitage Formation is variable in central Tennessee and adjacent regions. Near South Carthage (locality 48) the Millbrig lies at the contact between the Carters Limestone and the Hermitage Formation. Along Sequatchie Valley in Bledsoe, Sequatchie, and Marion Counties, however, Wilson (1949) noted that the T-4 (Millbrig) is overlain by 1 m to 2 m of Carters Limestone. Likewise, Haynes (1994) has shown that the Carters Limestone overlies the Millbrig K-bentonite southeast of central Tennessee at Davis Crossroads and Rising Fawn, Georgia (localities 50 and 51), and Fort Payne, Alabama (locality 52).

Near South Carthage (locality 48) a 15 cm thick K-bentonite is present in the uppermost Carters Limestone 0.6 m below the Millbrig. In an outcrop situated 4 km to the south, this bed is absent, but its position is marked by a prominent hardground. This may be the same bed that is widely present in this stratigraphic interval in eastern Kentucky and western Virginia.

Wilson (1949, p. 49) described K-bentonite T-5, present in the Hermitage Formation, as a unit that "sometimes consists of two beds separated by a foot or more of argillaceous limestone." The lithologic relationships and stratigraphic position suggest that this bed may be equivalent to the Capitol "Metabentonite" of Kentucky. The bed is exposed in roadcuts along U.S. Highway 70N near Carthage, Smith County, Tennessee (Wilson, 1949).

Southernmost Appalachians

This region includes the outcrop belts and subsurface of southeastern Tennessee, northeastern Alabama, and northwestern Georgia (Fig. 28). The K-bentonite beds range in age from mid-Whiterockian to late Cincinnatian (Fig. 29). Many Middle and Upper Ordovician rock units of central Tennessee are traceable southeastward through the subsurface into the outcrop belts in the Sequatchie Valley and into the Valley and Ridge Province of southeastern Tennessee, northwestern Georgia, and northeasternmost Alabama (Milici, 1969; Milici and Smith, 1969; Drahovzal and Neathery, 1971). Consequently, the stratigraphic nomenclature developed in central Tennessee is applied in some parts of the southern Appalachians. East of the Helena and Kingston thrust faults (Fig. 28) there are abrupt changes in Ordovician lithofacies. The following discussion focuses first on the stratigraphic succession west of the thrust faults and then on the sections within the thrust sheets.

Ordovician K-bentonite beds were described by Fox and Grant (1944) from Tennessee and adjacent states. The primary emphasis of their report was on southeastern Tennessee in the vicinity of Chickamauga Dam near Chattanooga, Tennessee, where fourteen beds ranging in age from Whiterockian to Cincinnatian were exposed during construction of the dam. Fox and Grant correlated K-bentonites in this region to beds in central and eastern Tennessee, northeast Alabama, and southwestern Virginia. The oldest bed, their B-1 ("B" for bentonite), is as much as 90 cm thick and is present near the base of the Murfreesboro Limestone in eastern Tennessee. Fetzer (1973) found B-1 to be associated with Midcontinent conodont Fauna 6 and suggested that it may be equivalent to a K-bentonite in the Dot Limestone in southwestern Virginia. By extrapolation from well-studied sections in Kentucky, the Murfreesboro Limestone is probably *P. aculeata* Zone in age (Sweet, 1984). Bed B-2 of Fox and Grant (1944), as much as 12 cm thick, occurs in the Carters Limestone and was correlated by Wilson (1949) with T-2 in central Tennessee.

The two thickest and most widespread beds described by Fox and Grant (1944), B-3 and B-6, are equivalent to the T-3 (Deicke) and T-4 (Millbrig) in central Tennessee. This correlation has been substantiated by recent outcrop studies (Milici and Smith, 1969; Drahovzal and Neathery, 1971; Haynes, 1994). Furthermore, both beds can be traced on wireline logs from the outcrop area in central Tennessee through the subsur-

Figure 31. Quarry in the Carters Limestone (C) and basal Hermitage Formation (H) near Shelbyville, Tennessee (locality 49). The Deicke K-bentonite (D), ~30 cm thick, forms a prominent reentrant on the high wall of the quarry. The Millbrig K-bentonite (M) is 80 cm thick and is situated at the contact between the Carters and Hermitage at the break in slope marked by the lowest cover of vegetation.

face of the Cumberland Plateau to outcrops in the Valley and Ridge region (Huff and Kolata, 1990).

In the Chattanooga area, the Deicke (B-3) is as much as 90 cm thick and is characterized by a basal 20 cm thick bed described by Fox and Grant (1944, p. 325) as "greenish-gray, sandy arkosic bentonite" overlying a 3 cm to 4 cm thick bed of "dense, black, fossiliferous chert." Beds B-4 and B-5 are approximately 3 cm thick and occur in the upper part of the Carters Limestone. The Millbrig K-bentonite (B-6), in the upper part of the Carters Limestone, is as much as 120 cm thick in the region of the Chickamauga Dam, making it one of the thickest occurrences in North America. Haynes (1994) showed that southwest of Chattanooga the Deicke and Millbrig can be correlated through outcrops at Davis Crossroads and Rising Fawn, Georgia, to Birmingham, Alabama (Fig. 32; localities 50, 51, 53). Northeast of Chattanooga they can be traced through eastern Tennessee into the Eggleston Formation in southwestern Virginia (Haynes, 1992).

Fox and Grant (1944) also noted the occurrence of three beds in the uppermost Carters Limestone (B-7, B-8, B-9), three beds in the Hermitage Formation (B-10, B-11, B-12), one in the Bigby-Cannon Limestone (B-13), and one in the Sequatchie Formation (B-14), all present in the Chattanooga area. Most of these beds, except for B-8 (30 cm to 70 cm), are relatively thin (1 cm to 12 cm) and their correlation beyond the Chattanooga region is uncertain. Haynes (1994) has noted several K-bentonite beds, other than the Deicke and Millbrig, in the upper part of the Carters Limestone and lower part of the Hermitage Limestones in northwestern Georgia and northeastern Alabama. These beds may be equivalent to those in the Chattanooga region.

Ordovician rocks in the thrust sheets bounded by the Helena and Kingston thrust faults were tectonically transported westward into juxtaposition with the Midcontinent platform facies. Presumably, the thrust sheets contain K-bentonite beds that were deposited nearer to the source volcanoes than those west of the thrust faults. The oldest reported K-bentonite bed in the region occurs in the lower part of the Athens Shale east of the Helena fault in northeastern Alabama (Drahovzal and Neathery, 1971). The bed is present near Calera, Alabama (Figs. 28 and 33; locality 54), in strata bearing conodonts characteristic of the lower *E. reclinatus* Subzone of the *P. serra* Zone (Schmidt, 1982). The base of the Athens Shale corresponds to a level in the lower *Hustedograptus teretiusculus* graptolite Zone (Finney, 1983).

Other K-bentonites occurring southeast of the Helena fault include two beds, 4.5 cm and 3.5 cm thick, in the Little Oak Limestone near Ragland, Alabama (Figs. 28 and 33; locality 55), reported by Drahovzal and Neathery (1971). Reexamination of this outcrop led Haynes and Melson (1995) to conclude that there is only one K-bentonite bed in the Little Oak and that thrust faulting has produced a duplicate stratigraphic section. The bed is believed to be in the *Pygodus anserinus* Zone (Fetzer, 1973; Schmidt, 1982; Kurapkat, 1986). Further, Drahovzal and Neathery (1971) report a single K-bentonite bed in the upper part of the Athens Shale in Shelby County, Alabama (locality 56). No conodonts were observed but graptolites suggest the likelihood of an early *Baltoniodus gerdae* Subzone age (Fetzer, 1973).

Drahovzal and Neathery (1971) also report the occurrence of two closely spaced K-bentonites in the upper part of the Colvin

Figure 32. Roadcut in the Chickamauga Limestone along the Red Mountain Expressway in Birmingham, Alabama (locality 53). Two prominent reentrants at center of image mark the position of thin Mohawkian age K-bentonite beds.

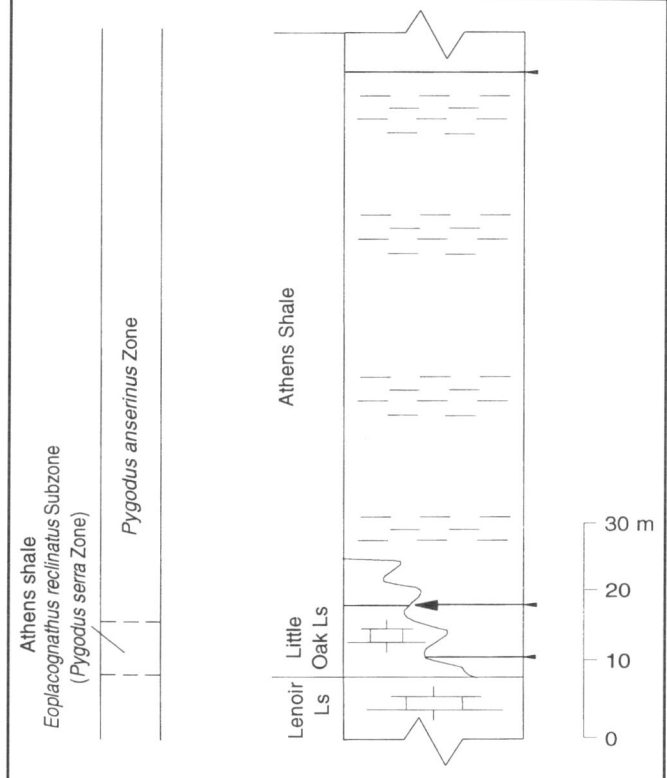

Figure 33. Composite stratigraphic column based on outcrops at Ragland, Calera, and Wilsonville, Alabama (localities 54, 55, and 56). All sections are situated southeast of the Helena Fault. K-bentonites indicated by arrow heads on right side of column. Abbreviation: limestone, Ls.

Mountain Sandstone at Greensport Gap, Alabama (Figs. 28 and 34; locality 57), the upper one of which they correlate with the Millbrig (T-4). Haynes (1994) observed only one K-bentonite at this locality, which he correlated with the Deicke based on abundant euhedral ilmenite grains. The lack of diagnostic fossils precludes determining an age of the bed based on biostratigraphy. At Alexander Gap, Alabama (Figs. 28 and 35; locality 58), the Deicke and Millbrig K-bentonites have been identified by Haynes (1994) in the Colvin Mountain Sandstone. Northeast of there the Colvin Mountain grades into the mixed siliciclastics of the Bays Formation extending northward into the eastern outcrop belt of southwestern Virginia. The Millbrig is locally absent in northwestern Georgia and eastern Tennessee but occurs more consistently in the Bays Formation from northeastern Tennessee to near Daleville in western Virginia (Haynes, 1994).

Southwestern Virginia and eastern Tennessee

Correlating K-bentonites within southwestern Virginia and eastern Tennessee is a challenge because of the structural and stratigraphic complexities in this part of the Valley and Ridge province. Generally, the Middle and Upper Ordovician rocks are characterized by significant along-strike facies changes making it difficult to determine the continuity of rock units and their age relations. Furthermore, thrust faulting has telescoped various lithofacies into juxtaposition far from their original depositional sites, making cross-strike correlation difficult. In spite of these difficulties, Rosenkrans (1936), Hergenroder (1966), and Haynes (1992, 1994) have shown that it is possible to correlate some K-bentonites along and across strike of the thrust sheets. Haynes's (1992, 1994) work is particularly important because it shows that some key beds can be identified on the basis of phenocryst mineralogy within different thrust sheets.

In southwestern Virginia and northeastern Tennessee, Ordovician rocks containing K-bentonites crop out in three distinct belts bounded by major thrust faults. Each belt is characterized by unique lithofacies (Kreisa, 1980; Carter and Chowns, 1986; Haynes, 1992) and biofacies (Jaanusson and Bergström, 1980). These are referred to as the western, central, and eastern outcrop belts (Fig. 36; Haynes, 1992). The western belt is the eastern extension of the Midcontinent platform facies, whereas the central and eastern belts are thrust sheets that were tectonically ramped, one over the other, during the late Paleozoic Alleghanian orogeny. Ordovician rocks in the eastern thrust sheet were deposited nearer to the cratonic margin than those in the western and central belts and thus presumably closer to the source volcanoes. An analysis of North Atlantic and Midcontinent conodonts indicates that the three outcrop belts also are characterized by distinct conodont faunas (Jaanusson and Bergström, 1980). These belts are the Lee Confacies (western) containing typical Midcontinent species, the Tazewell Confacies (central) containing North Atlantic and Midcontinent species, and the Blount Confacies (eastern) containing North Atlantic species.

Western outcrop belt. The stratigraphic succession noted by Miller and Fuller (1954) and Miller and Brosgé (1954) near Rose Hill, Lee County, southwestern Virginia, is representative of the western outcrop belt. Thirteen K-bentonite beds, each designated with an "R" for Rose Hill and ranging in age from early Mohawkian through Cincinnatian, were described by these authors (Fig. 37). The oldest K-bentonite in the region, R1, occurs in the Dot Limestone of *Plectodina aculeata* Zone age (Hall, 1986). Other beds in the western belt in Lee County, include R2 in the Hurricane Bridge Limestone, R3, R4, and R5 in the Hardy Creek Limestone, and R6 in the lower part of the Eggleston Formation. The age of the Hurricane Bridge is con-

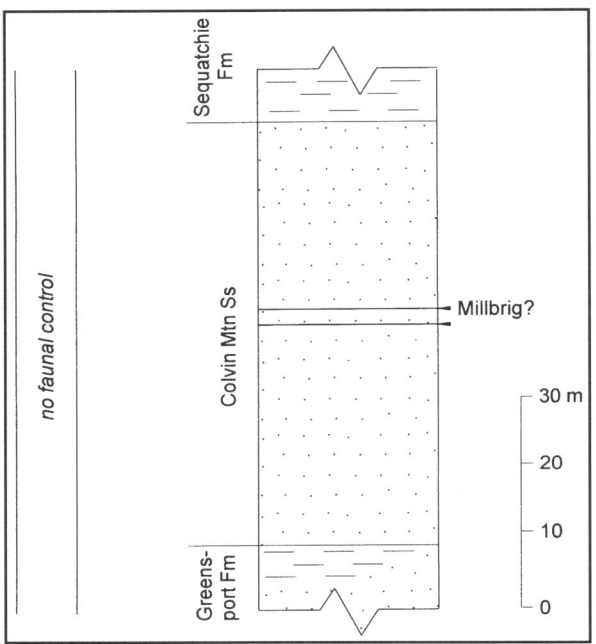

Figure 34. Generalized stratigraphic column of outcrop at Greensport Gap, Alabama (locality 57). Abbreviation: sandstone, Ss.

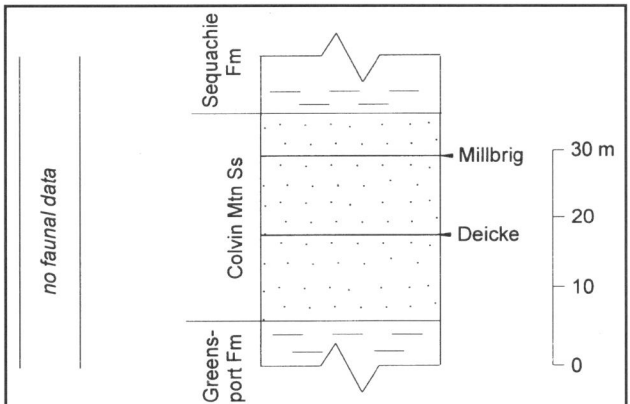

Figure 35. Generalized stratigraphic column of outcrop at Alexander Gap, Alabama (locality 58). Identification of Deicke and Millbrig based on phenocryst mineralogy (Haynes, 1994). Abbreviation: sandstone, Ss.

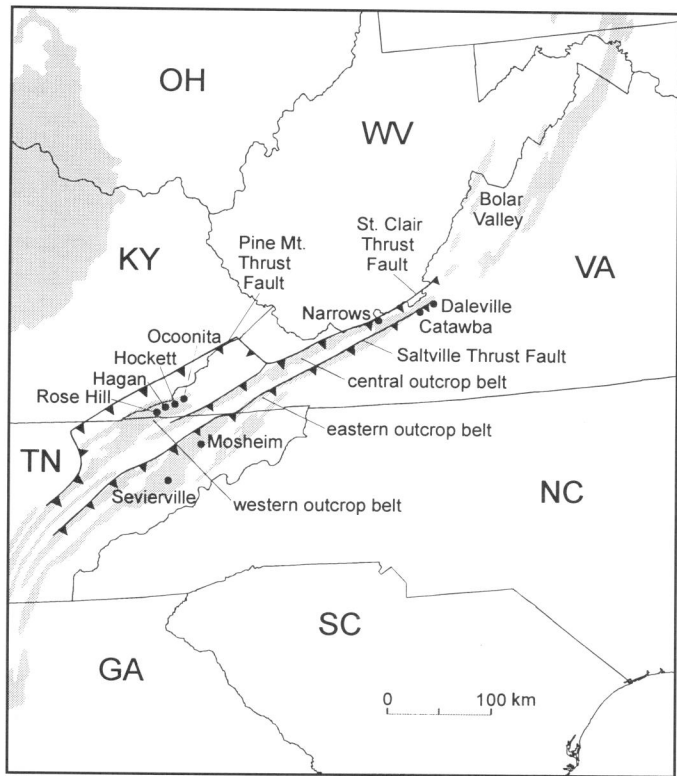

Figure 36. Ordovician outcrop belts (shaded) and major structural features of western Virginia and eastern Tennessee.

strained by the presence of *Erismodus quadridactylus* and the Hardy Creek by *B. compressa* (Bergström et al., 1988). The upper part of the Eggleston contains the thickest beds, R7 and R10, at 65 cm and 100 cm. These have been correlated with the Deicke and Millbrig (Haynes, 1992), are well exposed in outcrop at Hagan, Virginia (locality 59), and were encountered beneath the Pine Mountain thrust fault in the Gulf Oil No. 1 Price well in Russell County (locality 163). This interval lies within the *P. undatus* Chronozone (Fetzer, 1973; Hall, 1986; Bergström et al., 1988). Beds R11 and R12 occur in the Trenton Limestone and R13 in the Reedsville Shale. The upper part of the Trenton, containing R12, was assigned by Fetzer (1973) to Fauna 10 (*B. confluens* Chronozone). The K-bentonite R13 is probably early Cincinnatian in age.

Central outcrop belt. Fourteen K-bentonite beds, numbered V-1 through V-14 ("V" for Virginia), were described by Rosenkrans (1936) in the Tazewell County, Virginia, area of the central outcrop belt (Fig. 38; locality 60). Some of the more complete and well-exposed outcrops in the central belt have been described by Haynes (1992). Beds V-1 through V-3 occur in the Moccasin Formation and V-4 through V-11 in the Eggleston Formation. Recently, McVey (1993) observed an additional K-bentonite bed in the Moccasin Formation just above the V-3 in Tazewell County. Beds V-3 and V-4 correlate with the Deicke and Millbrig (Haynes, 1992). Bed V-6 was observed by

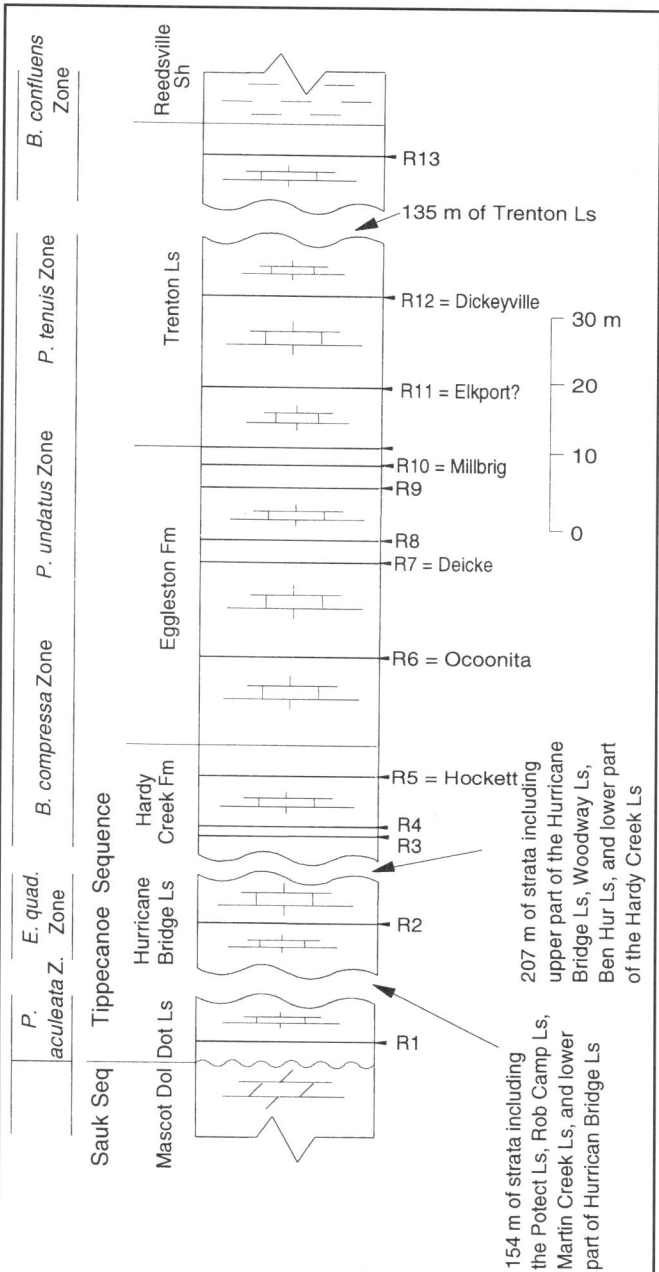

Figure 37. Generalized stratigraphic column of western Lee County, Virginia, typifying the western outcrop belt. K-bentonite symbols R1—R13 from Miller and Fuller (1954) and Miller and Brosgé (1954). Abbreviations: dolomite, Dol; limestone, Ls; shale, Sh.

Rosenkrans only at the Narrows, Virginia, section, and the stratigraphic interval was thought to be marked by a hiatus at localities where V-6 is absent. The bed is exposed at new localities near Bluefield and Tazewell, Virginia (John T. Haynes, 1994, personal communication). Petrological and geochemical analyses and additional field work are needed to sort out the stratigraphic relationships of V-6. Bed V-7 is thick (as much as

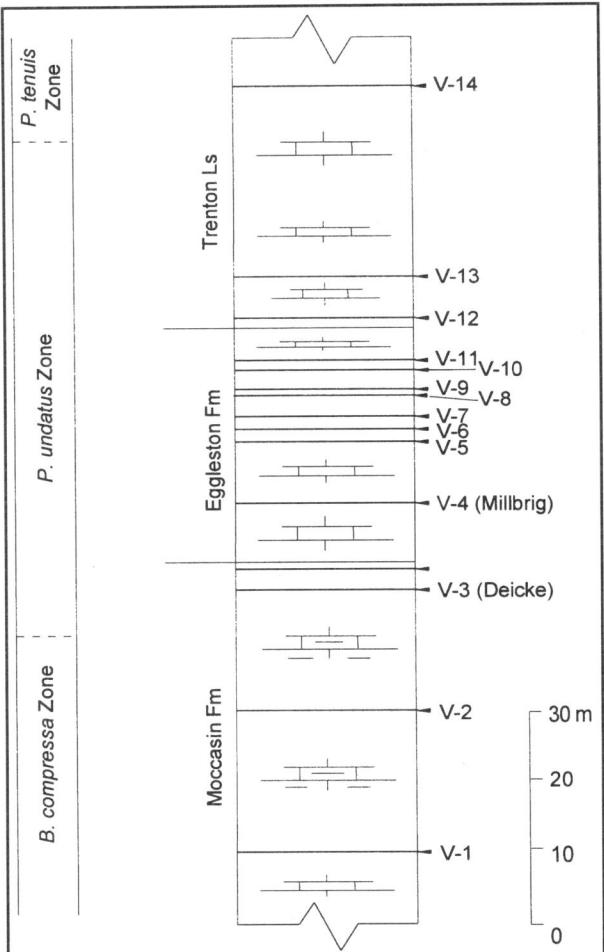

Figure 38. Generalized stratigraphic column of Tazewell County, Virginia, in the central outcrop belt. K-bentonite symbols V-1 through V-14 from Rosenkrans (1936). Abbreviation: limestone, Ls.

ian. One of the oldest K-bentonites is in a sinkhole in the Beekmantown Formation of the Knox Group on the sub-Tippecanoe unconformity (Fig. 39), uncovered during construction of Douglas Dam, 13 km north of Sevierville, Sevier County, Tennessee (Laurence, 1944). Stratigraphic relations suggest that the 15 m thick bed was deposited in a depression on the unconformity during mid-Whiterockian time prior to deposition of the regionally extensive Lenoir Limestone which contains the *P. serra* Zone fauna (Kurapkat, 1986). The bed likely represents the accumulation and concentration of volcanic ash in a sinkhole in the Beekmantown Formation prior to deposition of the Tippecanoe sequence.

Near Mosheim, Tennessee (Fig. 39; locality 61), five K-bentonite beds are present in the lower part of the Blockhouse Shale (Fetzer, 1973). Conodont distribution in this part of the Blockhouse indicates an age of late *P. serra* Zone, early *E. robustus* Subzone (Fetzer, 1973; Bergström, 1973; Kurapkat, 1986). A single K-bentonite was also reported by Fetzer (1973) from the *P. anserinus* Zone in the upper part of the Blockhouse at the Mosheim locality. Furthermore, a succession of K-bentonite beds nearly 2 m thick has been reported from the middle of the Blockhouse Shale, precise age unknown, at Boyds Creek approximately 30 km east of Knoxville, Tennessee (Ruppel and Walker, 1977).

Five K-bentonites, including the Millbrig and V-7, are pres-

100 cm) and widespread in the central belt. Haynes (1994) suggested that V-7 correlates with R-12 in the western belt at Hagan, Virginia, B-8 at Chattanooga, Tennessee, and T-5 in central Tennessee. Beds V-8 through V-11 occur within a thin stratigraphic interval characterized by blocky fractures that Rosenkrans (1936) referred to as the "cuneiform beds." Beds V-12 through V-14 are present in the Trenton Limestone.

Biostratigraphic control of the K-bentonite–bearing rocks in the central outcrop belt is relatively poor. Inferred faunal zones must be extrapolated from lithologically similar but not identical stratigraphic sections nearby. Thus, the lower part of the Moccasin Formation appears to be of *B. compressa* age based on conodont studies in Bolar Valley, Virginia. An analysis of conodonts from the Eggleston Limestone and lower part of the Trenton Limestone near Tazewell, Virginia (locality 60), indicates that these rocks contain the *P. undatus* fauna (Fetzer, 1973).

Eastern outcrop belt. In the eastern outcrop belt, K-bentonites range in age from Whiterockian through Mohawk-

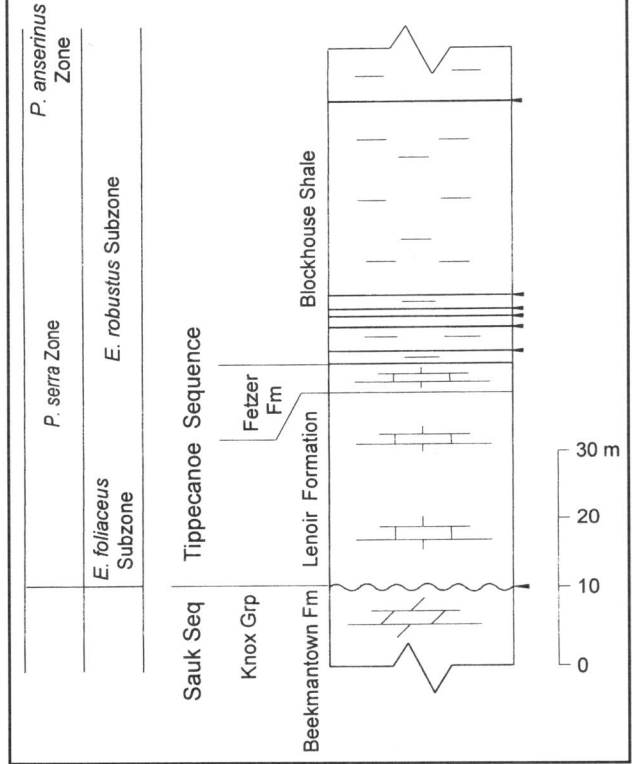

Figure 39. Generalized stratigraphic column of the eastern outcrop belt of western Virginia and eastern Tennessee showing unnamed K-bentonite beds.

ent in the Bays Formation near Catawba, Roanoke County, Virginia (Fig. 40; locality 62) (Haynes, 1992). Haynes suggested that the Deicke ash was deposited in the region of the eastern belt but because of reworking or removal by erosion, the ash was not preserved as a separate, discrete bed of bentonitic material in the Bays Formation.

Northern Virginia and eastern West Virginia

In northern Virginia, from Botetourt County northward to south-central Pennsylvania, Ordovician rocks containing K-bentonite beds are exposed in the Valley and Ridge province at intermittent outcrops in the Shenandoah Valley and in the western anticlines of Virginia and West Virginia (Fig. 41). The primary K-bentonite–bearing interval is in the Mohawkian carbonates (commonly classified as the Black River and Trenton Limestones) and their siliciclastic equivalents. In the Shenandoah Valley, the Bays Formation of the eastern outcrop belt grades northeastward into the Edinburg and Oranda Formations in northern Virginia (Cooper and Cooper, 1946) and into the Chambersburg Formation farther northeast in West Virginia, Maryland, and Pennsylvania. In the western outcrop belt of westernmost Virginia, the Eggleston and Trenton Limestones grade northeastward into the Nealmont and Dolly Ridge Limestones, respectively (Patchen et al., 1985). Stratigraphic sections tying the K-bentonite-bearing units of the southwestern region to northern Virginia are discussed by Rosenkrans (1936).

Key stratigraphic sections for the eastern outcrop belt are exposed in Shenandoah County, Virginia. The K-bentonite stratigraphy for this region was described by Kay (1935). At Tumbling Run southwest of Strasburg, Virginia, (Figs. 41, 42, and 43; locality 63) six K-bentonites occur in the Botetourt Member of the Edinburg Formation and one near the middle of the Edinburg (Rader and Read, 1989; John T. Haynes, 1993, personal communication). Nearby, on the north side of Strasburg (locality 64), at the type section for the Oranda Formation, three K-bentonites are present in the upper part of the Edinburg Formation, five beds in the overlying Oranda, and eight beds in the Martinsburg Formation (Cooper and Cooper, 1946; Rader and Read, 1989). Based on chemical fingerprinting data, McVey

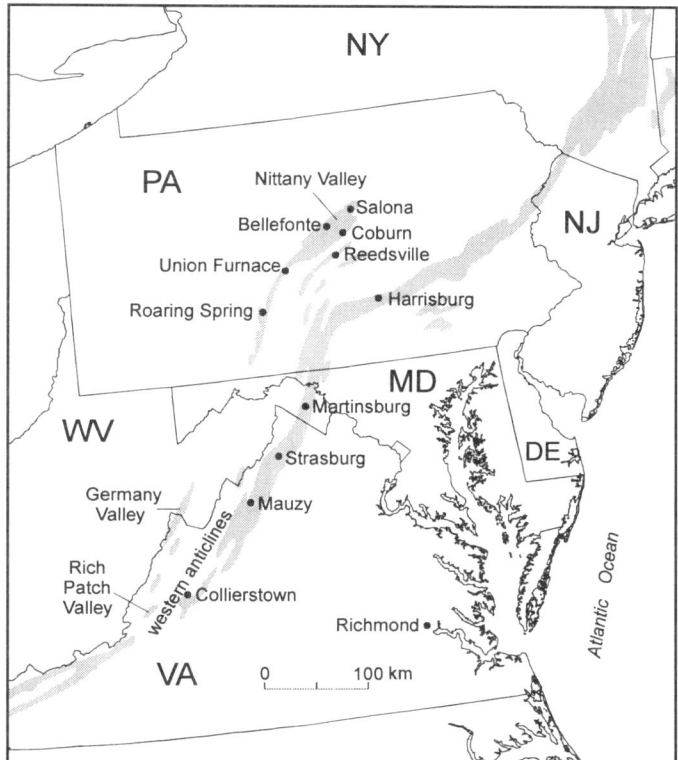

Figure 41. Ordovician outcrop belts (shaded) and major structural features in northern Virginia, eastern West Virginia, Maryland, and central Pennsylvania.

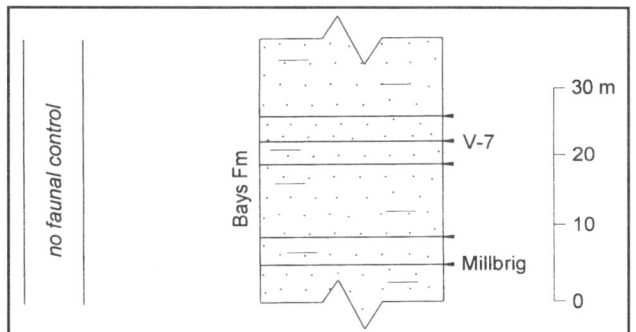

Figure 40. Generalized stratigraphic column based on outcrop at Catawba, Virginia (locality 62), in the eastern outcrop belt. Identification of Millbrig and V-7 based on phenocryst mineralogy (Haynes, 1994).

(1993) correlated the Deicke and Millbrig K-bentonites with beds 17 and 22 through 24, respectively, of Cooper and Cooper (1946). The Deicke is the uppermost K-bentonite in the Oranda, and the Millbrig is the basal bed in the overlying Martinsburg Formation. The lower Edinburg (Botetourt Member) occurs in the broad interval of the *Baltoniodus gerdae* Subzone, *Amorphognathus tvaerensis* Zone, and the rest of the Edinburg is higher in the *A. tvaerensis* Zone of the North Atlantic Conodont Province (Bergström, 1971; Fetzer, 1973). The lower Martinsburg Formation is of *Corynoides americanus* Zone age.

Farther south in the Shenandoah Valley near Collierstown, Virginia (locality 65), a single K-bentonite has been observed in the Edinburg Formation 56 m above the base (Cooper and Cooper, 1946; Fetzer, 1973). This bed is in the lower part of the *B. gerdae* Subzone, corresponding to the interval of Midcontinent Fauna 7 (Fetzer, 1973). Near Mauzy, Virginia (locality 66), four K-bentonites, including two beds interpreted by McVey (1993) to be the Deicke and Millbrig, occur in the Oranda Formation and two beds are present in the basal part of the Martinsburg Formation. Stratigraphic relations suggest that this is the same succession of K-bentonites, including the Deicke and Millbrig, that occurs in the Strasburg area discussed above. Rosenkrans (1934a) presented a bed-for-bed correlation of K-bentonites in the Oranda and Martinsburg Formations at the

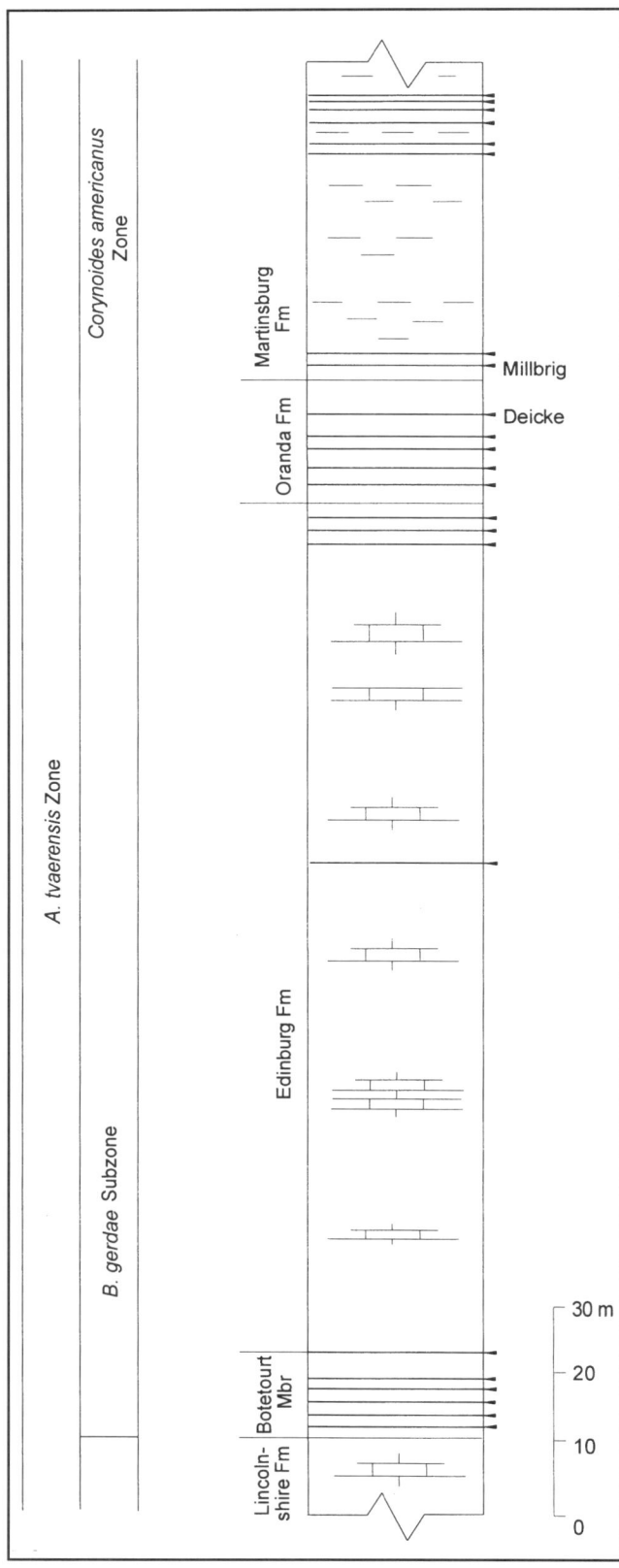

Figure 42. Generalized stratigraphic column based on outcrops at Tumbling Run and Strasburg, Virginia (localities 63 and 64). Correlation of the Deicke and Millbrig is based on chemical fingerprinting methods (McVey, 1993). Abbreviations: Formation, Fm; Member, Mbr.

Mauzy and Strasburg sections with the K-bentonite succession in the Salona Formation of central Pennsylvania. The correlation is supported by conodont data (Fetzer, 1973).

In the northern part of the eastern outcrop belt, carbonate rocks in the upper part of the Oranda Formation grade to a basinal shale facies of the Martinsburg Formation. These stratigraphic relationships are particularly well exposed in the quarry at Martinsburg, Berkeley County, West Virginia (locality 67). Here the Deicke and Millbrig have been identified in a succession of 12 K-bentonite beds at the base of the Martinsburg Formation (McVey, 1993). Faunal control is sparse, but stratigraphic relations suggest an early *C. americanus* Zone age for the Martinsburg Formation at this locality (O'Neill, 1985). Evidence of *P. undatus* has been found at the base of the Martinsburg at the Martinsburg quarry (Stephen A. Leslie, 1995, personal communication).

In west-central Virginia and eastern West Virginia, outcrops of Ordovician rocks are confined to five anticlinal valleys collectively referred to as the western anticlines (Fig. 41). All lie within the northern continuation of the western outcrop belt. K-bentonites are present in the Middle Ordovician Nealmont and Dolly Ridge Formations (Fig. 44). Biostratigraphic control is sparse in these formations because of the lack of diagnostic fossils in the shallow-water carbonates that characterize this stratigraphic succession. Based on extrapolation from the K-bentonite–bearing intervals in surrounding regions, the Nealmont/Dolly Ridge beds are probably *P. undatus*/*P. tenuis* Zone in age. Correlation of the Deicke and Millbrig K-bentonites from the southern Valley and Ridge into the southernmost anti-

Figure 43. Upper Oranda Formation (O) and lower Martinsburg Formation (Mb) exposed near Tumbling Run, Virginia (locality 63). Thin K-bentonite beds are marked by prominent reentrants. For detailed description of outcrop and K-bentonite stratigraphy see Cooper and Cooper (1946) and Rader and Read (1989). John T. Haynes stands at base of outcrop.

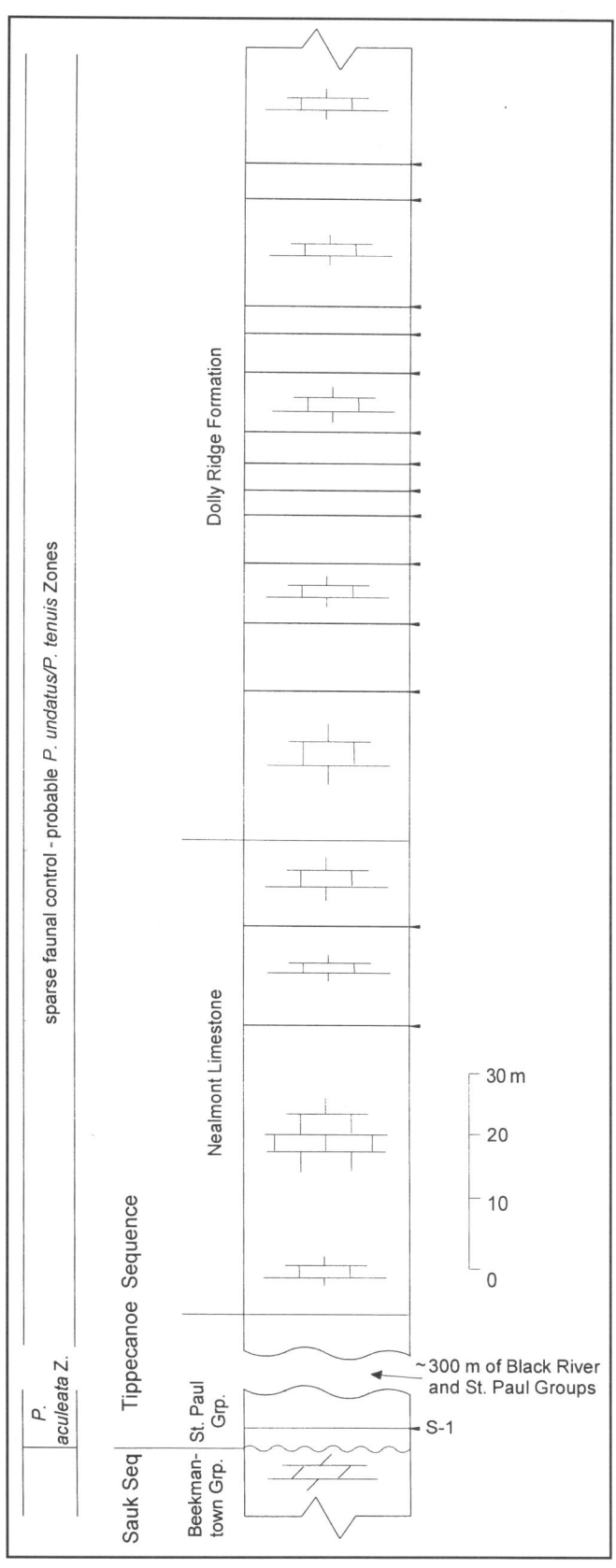

Figure 44. Generalized stratigraphic column of the western anticlines of west-central Virginia and eastern West Virginia showing unnamed K-bentonite beds. Abbreviations: Formation, Fm; Group, Grp; and Sequence, Seq.

cline, Rich Patch Valley, was proposed by Haynes (1992). Correlations of K-bentonite beds between outcrops in the western anticlines are not yet established; however, preliminary work suggests that in Germany Valley the Deicke and Millbrig are present in the lower Dolly Ridge Formation (John T. Haynes, 1993, personal communication). In Germany Valley, Perry (1972) recognized two K-bentonites in the Nealmont and twelve in the Dolly Ridge (Fig. 44). Correlation of the lower Dolly Ridge with the Salona Formation of the Nittany Valley in Pennsylvania is reasonably well established (Kay, 1956; Perry, 1972).

Thirteen K-bentonites, designated S-1 through S-13, were observed in drill cuttings and wireline logs from a well in Pendleton County, West Virginia (Perry, 1964). Bed S-1 is present at the contact between the Beekmantown and the overlying St. Paul Group. Faunal control is sparse, but stratigraphic relationships suggest a late Whiterockian, possibly *P. aculeata* Zone age. Beds S-2 through S-13 are present in the Nealmont and Edinburg Formations. The latter K-bentonites are probably the same cluster reported by Perry (1972) in outcrops of the Nealmont and Dolly Ridge Formations in Germany Valley of West Virginia. If so, they probably are of *P. undatus/P. tenuis* Zone age. Some of the beds have been correlated recently in a network of cross sections based on the deeper wells in Ohio, Pennsylvania, Virginia, and West Virginia (Ryder, 1991, 1992a, 1992b).

Central Pennsylvania

Ordovician rocks in central Pennsylvania crop out in linear belts along the margins of the eroded anticlinal valleys including the Nittany Valley (Fig. 41). Twenty-nine K-bentonite beds have been described from Mohawkian rocks in this region (Fig. 45; Bonine and Honess, 1929; Whitcomb, 1932; Rosenkrans, 1934a; Kay, 1944; Thompson, 1963; Rones, 1969; Cullen-Lollis and Huff, 1986).

The oldest known K-bentonites are in the upper part of the Snyder Formation of *B. compressa* Zone or slightly older age. Three beds, between 1 cm and 2 cm thick, are present in the Snyder at Union Furnace (locality 68) according to Smith et al. (1986). One of the beds in the upper Snyder was correlated by Rones (1969) between Pleasant Gap and Oak Hall, Pennsylvania (localities 69 and 70). The overlying Linden Hall Formation contains at least six locally prominent K-bentonites, as much as 25 cm thick, designated "A" through "F" by Rosenkrans (1934a). Three additional beds at the top of the Linden Hall and basal part of the overlying Nealmont Formation were recognized by Rones (1969) and designated N_1 through N_3. Correlation of the K-bentonites shows the diachroneity of stratigraphic members within the Linden Hall and unconformable relations at the top of the formation (Rones, 1969). Three or four locally

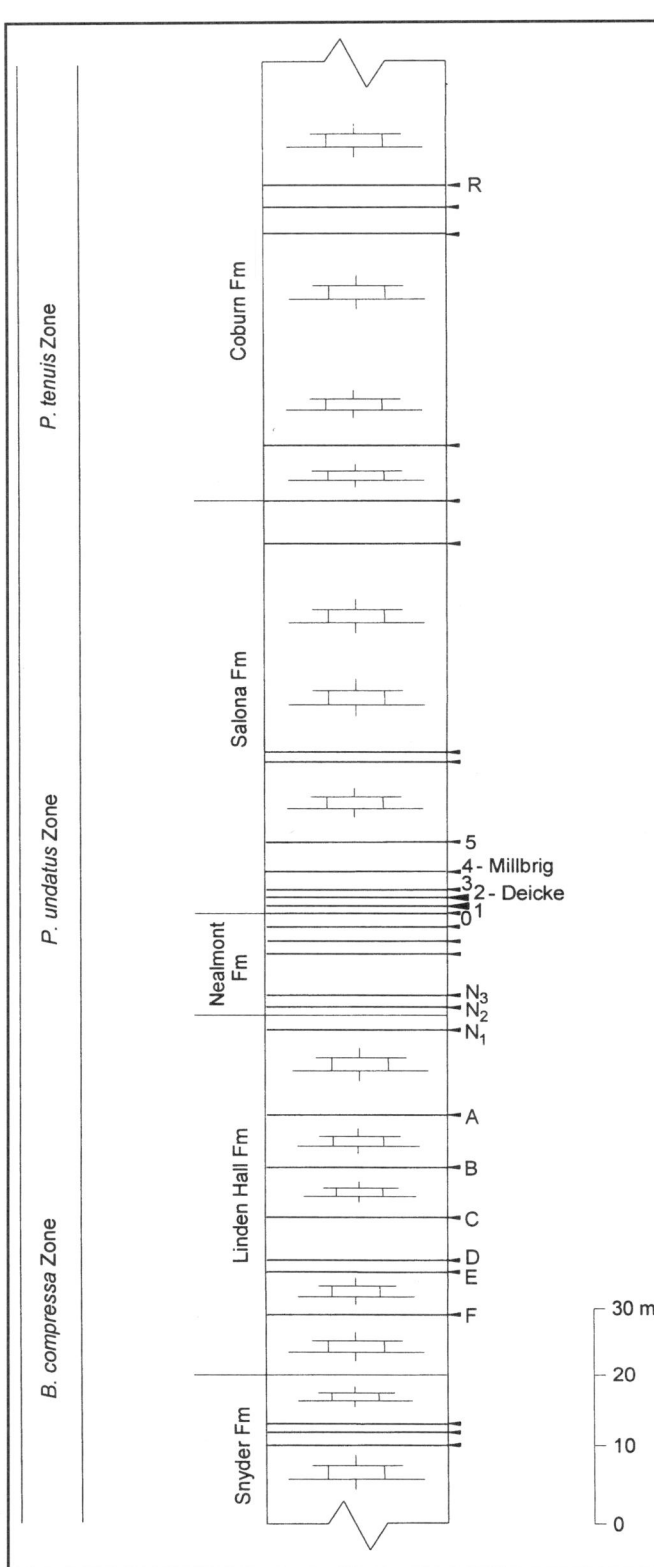

Figure 45. Generalized stratigraphic column of central Pennsylvania. K-bentonites symbols A–F and 0–5 from Rosenkrans (1934a), N_1–N_3 from Rones (1969). Correlation of the Millbrig and Deicke is based on chemical fingerprinting methods (McVey, 1993). Abbreviation: Formation, Fm.

occurring K-bentonites are present in the upper part of the Nealmont Formation at Union Furnace (Smith et al., 1986).

The Salona Formation, overlying the Nealmont, contains six persistent K-bentonite beds as much as 25 cm thick in the basal part of the formation (Fig. 46). These are designated, in ascending order, No. 0 through No. 5 (Whitcomb, 1932; Rosenkrans, 1934a). Beds No. 0 through No. 4 have unique and identifiable chemical signatures that can be recognized in the central Pennsylvania outcrop belt (Cullen-Lollis and Huff, 1986). The Deicke and Millbrig have been chemically correlated with beds No. 2 and No. 4, respectively (McVey, 1993). Several K-bentonites have been observed in the upper part of the Salona and lower part of the Coburn. The so-called "R" bed, one of the more persistent beds in the region, has been shown to occur in the lower part of the Coburn Formation near Salona, Bellefonte, and Roaring Springs, but south and east of there near Coburn, Reedsville, and Shade Gap, it is found in the upper part of the Salona Formation (Thompson, 1963), thus showing the diachronous nature of these formations.

The upper half of the Linden Hall through the upper part of the Salona Formation contains the *P. undatus* fauna (Leslie and Bergström, 1995a), and the upper Salona and overlying Coburn Formation belong to the *P. tenuis* Zone (Sweet, 1984).

Southeast of the Nittany Valley, K-bentonite–bearing Middle Ordovician rocks occur in a belt of disconnected exposures from Harrisburg, Pennsylvania, northeastward into New Jersey (Figs. 41 and 47). In the Harrisburg area, this limestone-dominated succession is assigned to the Myerstown and Hershey Formations (Prouty, 1959) and in easternmost Pennsylvania and northwestern New Jersey it is assigned to the Jacksonburg Formation (Spencer et al., 1908; Miller, 1937; Sherwood, 1964). Four beds ranging in thickness from 4 cm to 33 cm are present in the Meyerstown Formation at Nazareth, Pennsylvania (Miller, 1937). The thicker bed, which has been observed at several localities, is believed to be equivalent to bentonite 2 (Rosenkrans, 1934a) in the Salona Formation of the Nittany Valley (Prouty, 1959). By our correlations this would be the Deicke K-bentonite Bed.

Conodonts of the Jacksonburg Formation are characteristic of the North American Midcontinent conodont province and have been correlated with faunas of the *P. undatus* Chronozone, with particular similarity to the faunas of the Decorah Shale of Minnesota and the Kimmswick Limestone of eastern Missouri (Barnett, 1965).

New York

Ordovician K-bentonite beds are exposed in the Mohawk and Black River Valleys in central and northern New York State

Figure 46. Vertical beds of limestone and a 30 cm thick K-bentonite bed in the Salona Formation in the New Enterprise Quarry southeast of Tyrone, Pennsylvania (locality 68). Joan Cullen-Lollis is measuring the section.

Figure 47. K-bentonite beds (light colored beds) in the Oranda Formation exposed in the Pennsy Supply Quarry in Harrisburg, Pennsylvania (locality 221). The exposed stratigraphic succession is ~10 m thick and has been metamorphosed to phrenite grade (~300 °C) that altered the K-bentonites to a hard, dense consistency.

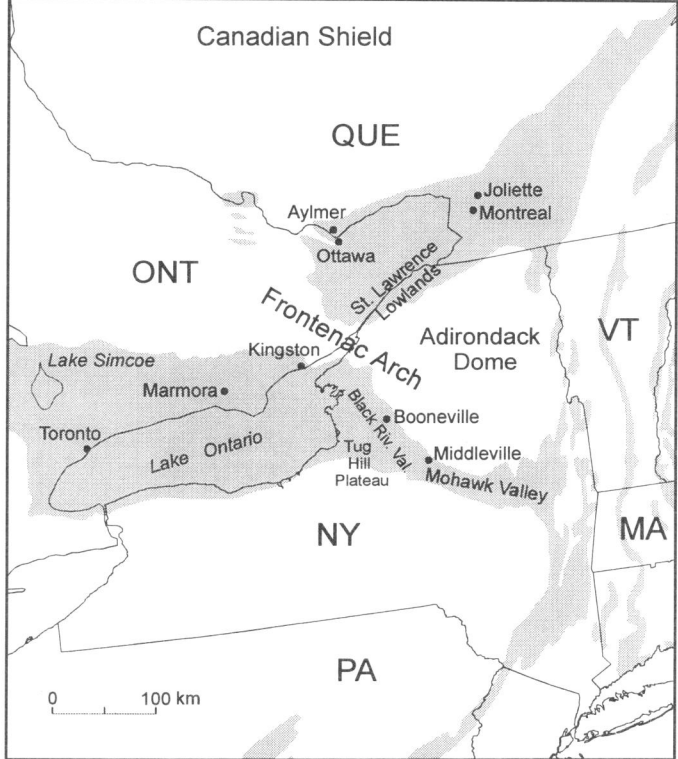

Figure 48. Ordovician outcrop belts (shaded) and major structural features in New York, southern Ontario and Quebec.

(Fig. 48). Known beds are confined to the Mohawkian Series and basal part of the Cincinnatian Series (Kay, 1935, 1937; Cisne and Rabe, 1978; Cisne et al., 1982; Cisne and Chandlee, 1982; Riva, 1972; Walker, 1973; Mitchell et al., 1994). The stratigraphic succession consists of carbonate rocks of the Black River Group overlain by carbonate and/or siliciclastic rocks of the Trenton Limestone/Utica Shale. Rocks within this interval were deposited across a shelf-slope-basin tectonic setting characterized by complex facies relations. There is no clear agreement in the literature regarding the litho- or chronostratigraphy of these rocks. Lithostratigraphy discussed herein follows Fisher (1977, 1982).

The upper part of the Lowville Formation, of *B. compressa* Zone age (Sweet, 1984), contains a K-bentonite bed (Fig. 49) present at four localities in the Tug Hill Plateau and Black River Valley region (Walker, 1973). The overlying Watertown Limestone in the Black River Valley contains the "Hounsfield metabentonite" (Kay, 1931), the type section being near Dexter, New York (locality 72). Kay correlated the Hounsfield widely in eastern North America, believing that the bed represented a unique event. Accordingly, Kay correlated the Hounsfield with the Millbrig K-bentonite in the Spechts Ferry Formation of the Upper Mississippi Valley region. Whitcomb (1932) showed that

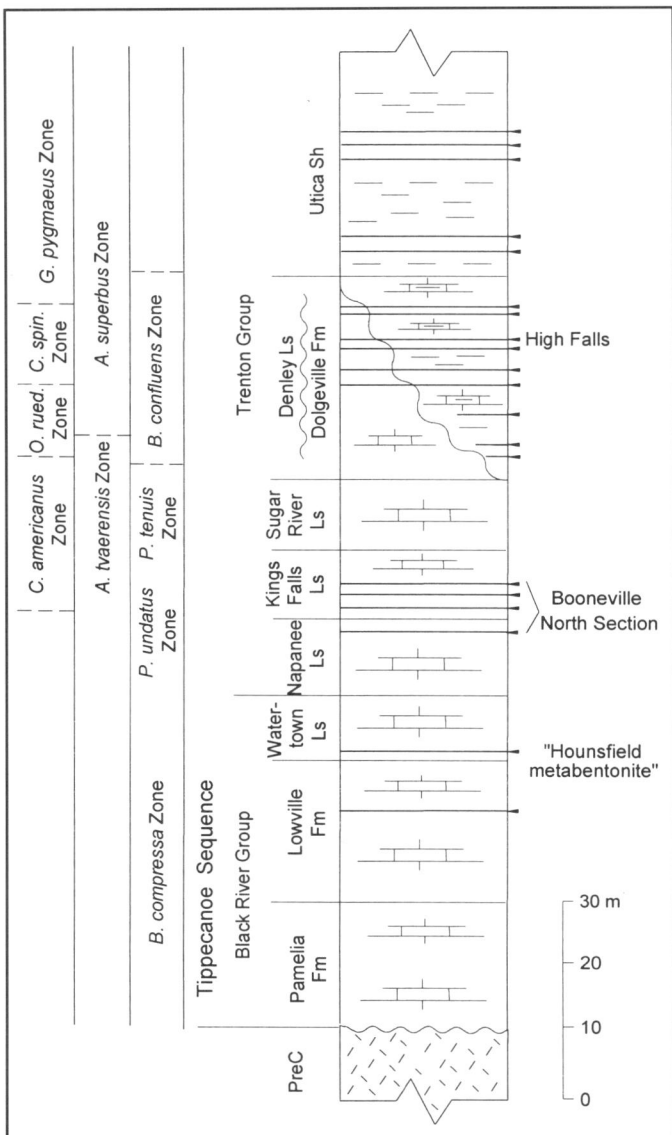

Figure 49. Generalized stratigraphic column of the western shelf facies in the upper Mohawk and Black River Valleys, New York. Abbreviations: Formation, Fm; limestone, Ls; shale, Sh.

there are at least six Middle Ordovician K-bentonite beds in central Pennsylvania, pointing out that Kay erred in assuming that only one bed is present in eastern North America. The Hounsfield cannot now be correlated confidently to stratigraphic sections outside of the type region. Strata containing the Hounsfield are of *P. undatus* Zone age (Sweet, 1984).

Near Boonville, New York (locality 71), a 2 cm to 4 cm K-bentonite occurs in the upper meter of the Napanee Limestone. The lower 3 m of the overlying Kings Falls Limestone contain three K-bentonites, each 2 cm to 5 cm thick. These K-bentonites are in approximately the same stratigraphic position as two 5 cm thick beds in the basal part of the Kings Falls Limestone near Middleville, New York (locality 73), and a single bed 2 m above the base of the Kings Falls at Buttermilk Creek near Middleville, New York (locality 74). The Napanee and basal Kings Falls lie within the *P. undatus* Midcontinent Chronozone (Goldman et al., 1994).

In the eastern part of the Mohawk Valley, most of the Mohawkian Series consists of the Utica Shale, a basinal, graptolite-bearing shale (Ruedemann, 1912; Fisher, 1977). The lower part of the Utica grades westward into an interbedded shale/limestone slope facies of the Dolgeville Formation, which grades westward into platform carbonates of the Denley Limestone of the Trenton Group (Fig. 50; Kay, 1937, 1953; Cisne et al., 1982; Goldman et al., 1994; and Mitchell et al., 1994). Cisne and Rabe (1978) and Cisne et al. (1982) utilized K-bentonites to construct a chronostratigraphic framework applicable to both facies in the Mohawk Valley. Recently, this framework has been challenged by Goldman et al. (1994) and Mitchell et al. (1994), who suggest miscorrelations of as large as 60 m based on integration of K-bentonite and biostratigraphic data. The new corre-

Figure 50. Trenton Group and Utica Shale exposed at Trenton Falls, New York (locality 222). Prominent reentrants in the Trenton Group limestones near the top of the falls mark the position of K-bentonite beds.

lations are supported by geochemical fingerprinting of glass inclusions within volcanic quartz phenocrysts from selected K-bentonite beds. The methods permit unambiguous matching of compositionally unique ash layers.

Approximately 60 K-bentonite beds, most of which are less than 5 cm thick, are present in the Utica Shale (Charles E. Mitchell, 1994, written communication). The greatest number of beds occur in the late Mohawkian *Corynoides americanus*, *Orthograptus ruedemanni*, and *Climacograptus spiniferus* North American graptolite Zones (Fig. 51). Approximately six to ten thin K-bentonites also occur in the overlying *Geniculograptus pygmaeus* Zone of the Utica Shale (Charles E. Mitchell, 1994, written communication). A small number of K-bentonite beds have been correlated westward from the Utica Shale to the carbonate platform. Six beds are believed to extend from the Dolgeville Formation westward into the Denley Limestone (Goldman et al., 1994) where they occur in the *Amorphognathus tvaerensis* and *A. superbus* North Atlantic conodont Zones (Fig. 51; Bergström and Mitchell, 1994). One of the thicker and more widespread beds, formally named the High Falls K-bentonite Bed, has been correlated on the basis of chemical fingerprinting from its type section at Trenton Falls eastward through the basin margin and slope facies (Mitchell et al., 1994). The regional extent of this bed is unclear.

Southern Ontario

This region includes Ordovician rocks in outcrop and subsurface extending westward from the Frontenac Arch at Kingston to Windsor and northward to and including Manitoulin Island (Figs. 48 and 52). The area encompasses the western St. Lawrence Lowland. K-bentonites have been observed in outcrops and in core and drill-cutting samples. The stratigraphic framework used here follows Liberty (1969) for the outcrop belt and Sanford (1961) for the subsurface. Little success has been achieved in relating the subdivisions of the Black River and Trenton Groups in outcrop and subsurface in southern Ontario (Derek K. Armstrong and Terry Carter, 1994, personal communications). Consequently, the outcrop and subsurface regions are addressed separately in the following discussion.

The Gull River Formation of southern Ontario, which is approximately equivalent to the Lowville of New York, contains two persistent K-bentonites ("MX" and "MH") described by Liberty (1969) in outcrop (Figs. 53 and 54). The Gull River contains the *B. compressa* conodont fauna (Sweet, 1984) characteristic of Midcontinent Fauna 7 (Barnes et al., 1981). Bed MX is approximately 3 cm thick and occurs at the top of the "Lower Member (Unit A)" of the Gull River, 7 m to 12 m below the base of the overlying Bobcaygeon Formation. The MH K-bentonite, commonly 3 cm thick but as much as 20 cm thick in the Burnt River vicinity (Liberty, 1969), is present in the upper part of the "Middle Member (Unit B)" of the Gull River Formation about 3 m below the base of the overlying Bobcaygeon Formation. Both MX and MH are exposed in the Coldwater Quarry northwest of Lake Simcoe (Fig. 55; locality

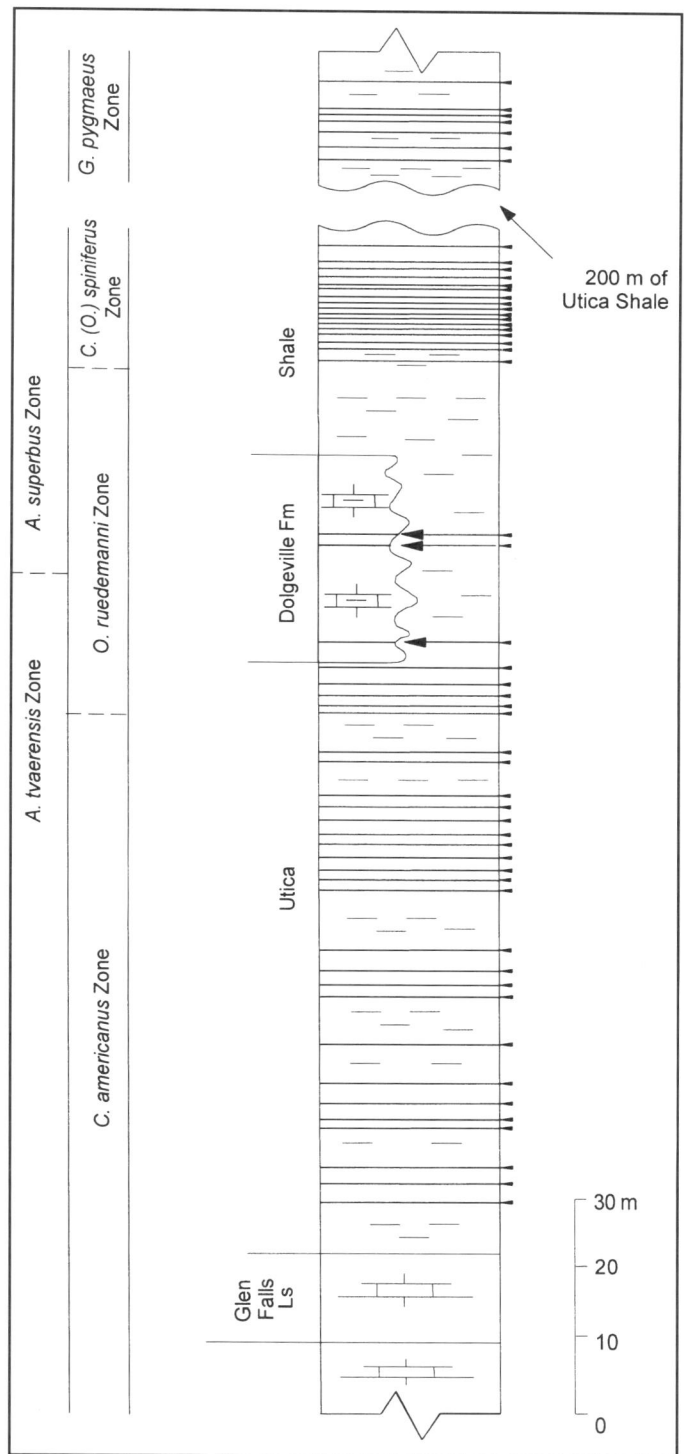

Figure 51. Generalized stratigraphic column of the basin and slope facies of the Mohawk Valley of New York. Abbreviation: limestone, Ls.

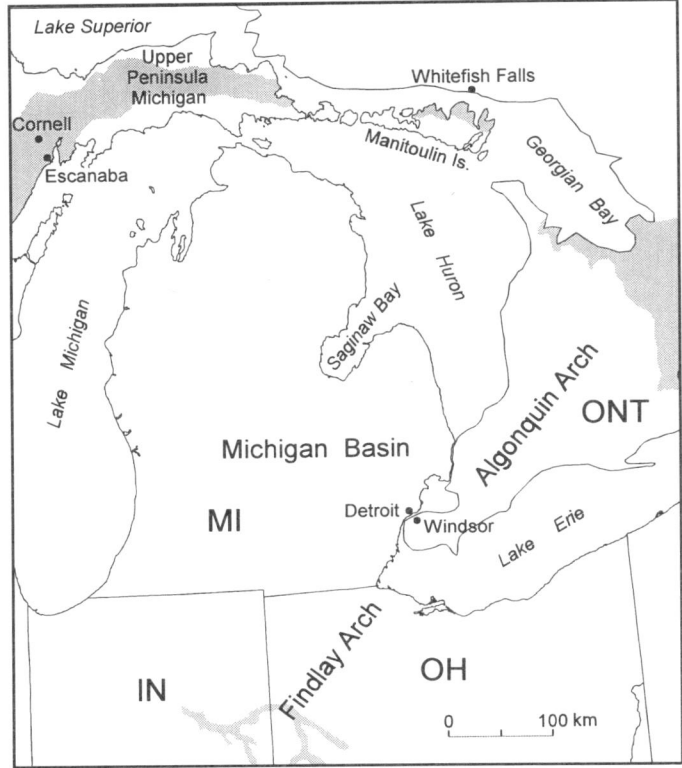

Figure 52. Ordovician outcrop belts (shaded) and major structural features around periphery of the Michigan Basin.

Figure 54. Roadcut through the Gull River Formation at Marmora, Ontario (locality 76). A 5 cm thick K-bentonite is marked by a prominent bedding plane near the middle of the outcrop.

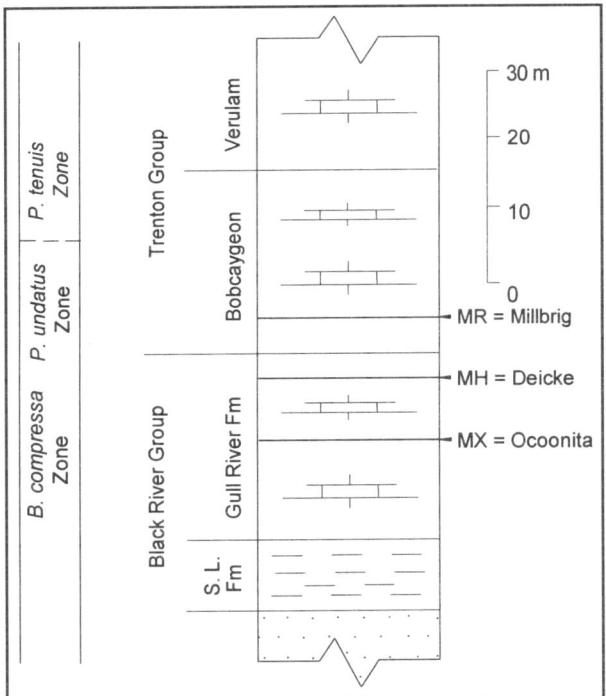

Figure 53. Generalized stratigraphic column of the outcrop region of southwestern Ontario. K-bentonite symbols MX, MH, and MR from Liberty (1969). Abbreviation: Shadow Lake Formation, S.L. Fm.

Figure 55. Gull River Formation exposed near Coldwater, Ontario (locality 75). Base of hammer marks position of thin K-bentonite bed.

75) and in a roadcut along Provincial Highway 7 on the west side of Marmora (locality 76).

The Bobcaygeon Formation contains a K-bentonite at the base of the "Middle Member (Unit D)" about 6 m above the base of the formation. The bed is referred to as MR (Liberty, 1969). Diagnostic conodonts in the Bobcaygeon succession include *P. undatus* in the lower part and *P. tenuis* in the upper part (Votaw, 1971, 1980; Bergström and Mitchell, 1992).

Several K-bentonites can be traced widely through the subsurface on wireline logs and drilling samples in southwestern Ontario (Sanford, 1961; Trevail et al., 1989; Trevail, 1988, 1990; Kolata et al., 1990). The oldest reported K-bentonite is present in subsurface samples at the base of the Gull River Formation, generally east of Toronto, Ontario (Sanford, 1961; Trevail, 1990). By extrapolation from the outcrop belt of southern Ontario, the lower Gull River appears to be of *B. compressa* Chronozone age.

Two relatively thick and persistent beds, identified as the Deicke and Millbrig (Trevail, 1990), occur near the contact between the Black River and Trenton Groups (Fig. 56). Correlation of these beds is supported by chemical fingerprinting data. The beds produce a prominent deflection on gamma ray and neutron logs where they occur in relatively pure carbonate rocks. By extrapolation from zoned sections in Ohio, both K-bentonite beds appear to be in the *P. undatus* Chronozone. The Deicke typically is situated about 3 m to 4 m below the contact between the Coboconk Formation and the overlying Kirkfield Formation. This interval increases to 8 m in the subsurface near Windsor, Ontario (locality 140), and as much as 11 m in the Humble Oil and Refining Co. No. 1 Hoppinthal well in Sanilac County, Michigan (locality 138). The thickest interval is in eastern Michigan and southwesternmost Ontario, site of the Michigan Basin depocenter during Middle Ordovician time (Catacosinos et al., 1991). We believe that the Deicke is equivalent to the "Black River shale" noted by drillers in the Michigan Basin (Lilienthal, 1978). The MH K-bentonite of Liberty (1969) appears to be equivalent to the Deicke.

The Millbrig is identified as the K-bentonite that typically occurs in the lower part of the Kirkfield Formation about 2 m to 4 m above the Coboconk-Kirkfield contact (Trevail, 1990). The geophysical signature of the bed is commonly obscure on wireline logs because of the inherent shaliness of the basal Kirkfield rocks. Trevail (1990) tentatively identified the Millbrig in a core sample from the Ontario Geological Survey 82-2, Harwich 25-I ECR drillhole in Kent County, Ontario (locality 207). The Millbrig occurs in the same stratigraphic interval as bed MR, described by Liberty (1969) from near the base of the Bobcaygeon Formation in outcrops in the Lake Simcoe region, southwestern Ontario.

The Sherman Falls Formation, which overlies the Kirkfield Formation, contains a persistent K-bentonite that has been observed on wireline logs (Plate 3) and in drilling samples in southwestern Ontario (Robert A. Trevail, 1993, personal communication). The bed typically occurs between 30 m and 40 m

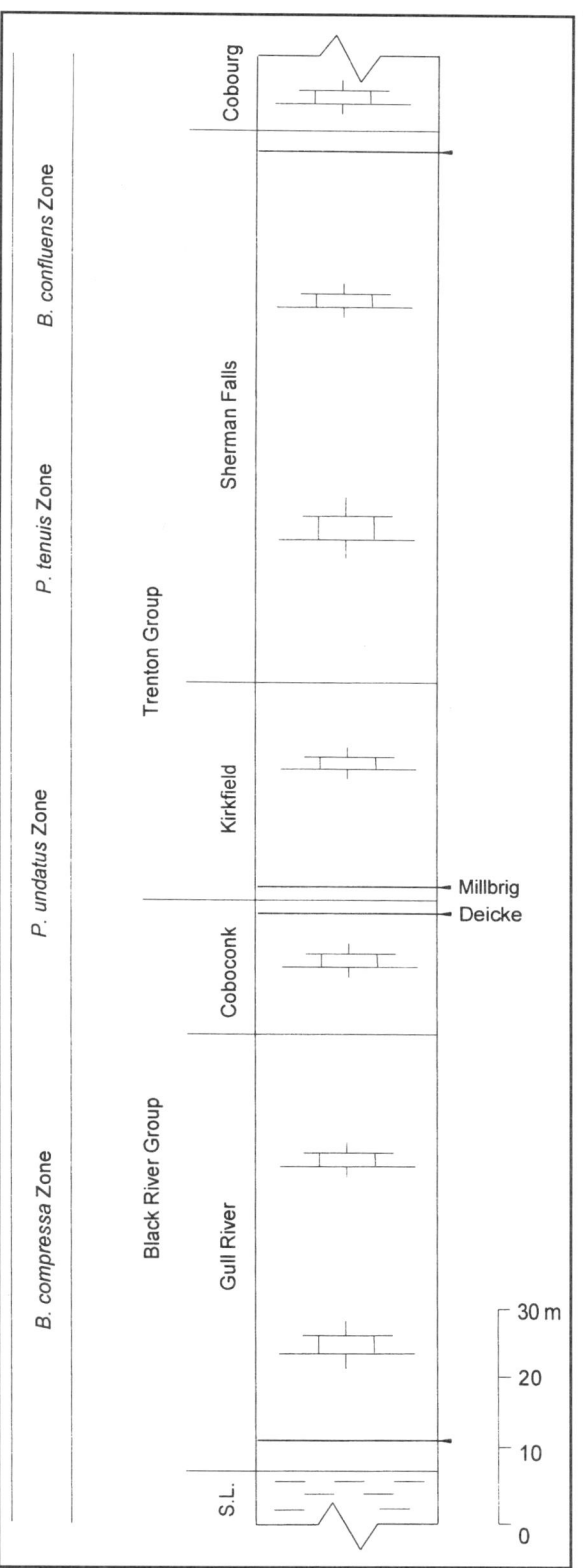

Figure 56. Generalized stratigraphic column of the subsurface region of southwestern Ontario. Abbreviation: Shadow Lake Formation, S.L.

below the Utica Shale in southwesternmost Ontario. It is 36 m below the Utica in Essex County (locality 140) and 39 m below the Utica in Oxford County (locality 143). East of Oxford County the bed is difficult to distinguish on wireline logs. The bed can be traced westward into the Michigan Basin where it is 45 m below the Utica in Wayne County, Michigan (locality 139), and 49 m below the Utica in Kalamazoo County (locality 134). Based on correlations of Barnes et al. (1981) the Sherman Falls is *B. confluens* Zone in age.

On Manitoulin Island (Fig. 52), two K-bentonites are exposed in a roadcut on Provincial Highway 68, 6.4 km south of Whitefish Falls (locality 77), both within the Swift Current Beds (Turinian Stage), a lateral equivalent of the Gull River Formation (Liberty and Shelden, 1968). The lowermost bed is about 3.3 m below the Swift Current/Cloche Island Beds (Figs. 57 and 58) contact and is 3 cm to 5 cm thick (Stephen A. Leslie, 1994, personal communication). The uppermost K-bentonite is about 7 cm thick and occurs about 37 cm below the top of the Swift Current. This same K-bentonite apparently is exposed in a road cut on Highway 68 about 12 km south of Whitefish Falls (locality 78) where it crops out 15 cm below the Swift Current/Cloche Island contact. A K-bentonite bed also was encountered in the Swift Current Beds (Black River Group) in the Manitoulin Island DDH No. 2 Elizabeth Bay drillhole (locality 208) on Manitoulin Island (Trevail, 1990). Although occurring in the same stratigraphic interval, it is unclear how the individual Manitoulin K-bentonites correlate with those in southwestern Ontario and the Michigan Basin. Conodont biostratigraphy indicates that the Swift Current beds are equivalent to the Gull River (Midcontinent conodont Fauna 7, *B. compressa* Chronozone) and the Cloche Island beds are approximately equivalent to the Bobcaygeon (conodont Fauna 8, *P. undatus* Chronozone) of southern Ontario (Barnes et al., 1978).

Figure 58. Roadcut near the Roosevelt Memorial on Manitoulin Island (locality 77). Relatively pure limestone of the Gull River Formation (GR) is overlain at the break in slope by argillaceous and shaly limestone of the Bobcaygeon Formation (B). A 5-cm-thick K-bentonite present near the top of the Gull River is marked by a prominent reentrant.

Michigan Basin

Ordovician rocks in the Michigan Basin are confined to the subsurface except for a narrow outcrop belt along the northern margin of the basin in the Upper Peninsula of Michigan near the town of Escanaba (Figs. 52 and 59). The oldest known K-bentonites in the Escanaba region occur in the Turinian Bony Falls beds (Black River Group) at Bony Falls on the Escanaba River (locality 79; Hussey, 1952; Stephen A. Leslie, 1994, personal communication). Approximately 12 m of argillaceous limestone are exposed along the river banks. Two K-bentonites as much as 2 cm to 3 cm thick occur in the upper 4 m of the

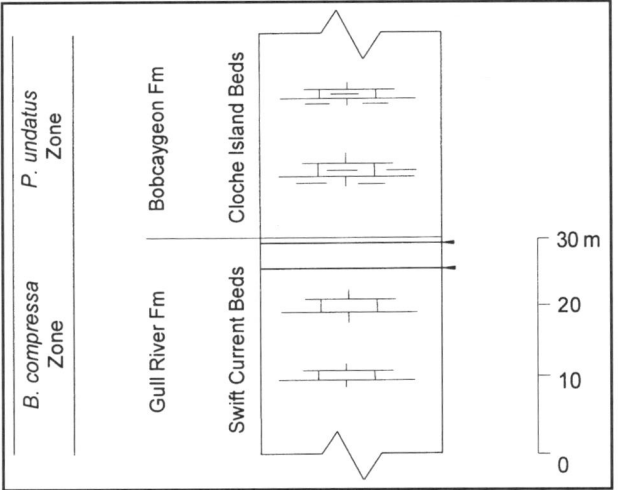

Figure 57. Generalized stratigraphic column of Manitoulin Island. Abbreviation: Formation, Fm.

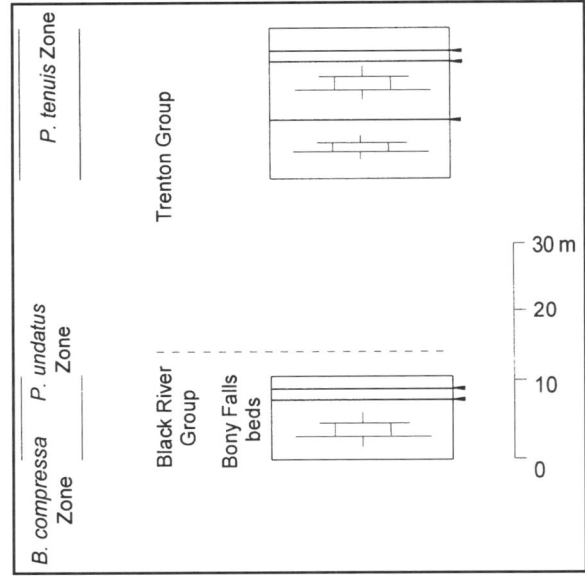

Figure 59. Generalized stratigraphic column of Black River and Trenton strata in the region of Escanaba, Michigan.

outcrop separated by 90 cm of crinoidal limestone (Fig. 60). Votaw (1980) assigned the Bony Falls to Midcontinent conodont Fauna 7 (*B. compressa* Chronozone).

Between Cornell and Escanaba, K-bentonites also have been reported from the Mohawkian Trenton Group (Fig. 59) in outcrops along the Escanaba River (Kay, 1935; Templeton and Willman, 1963). The composite section consists of approximately 20 m of argillaceous limestone. A K-bentonite in the Trenton Group exposed near Cornell (locality 80) was correlated by Templeton and Willman (1963) with the Nasset K-bentonite Bed (Willman and Kolata, 1978) in the Sherwood Member of the Dunleith Formation of the Upper Mississippi Valley. Two K-bentonites occurring higher in the Trenton (Sherman Falls Formation in Kay, 1935) near Chandler Falls (locality 81) were judged by Templeton and Willman (1963) to be equivalent to two beds in the Sinsinawa Member of the Wise Lake Formation exposed near Galena, Illinois, one of which (Dygerts K-bentonite Bed of Willman and Kolata, 1978) is widespread in the Upper Mississippi Valley. This correlation is questionable, however, because Votaw (1980) judged the Chandler Falls section to be in the *P. tenuis* Chronozone. Templeton and Willman's correlations must be regarded as tentative because a comprehensive biostratigraphic framework has not been established for either the Michigan Upper Peninsula or the Upper Mississippi Valley region.

Within the Michigan Basin, Ordovician rocks have been penetrated by many widely distributed oil and gas exploratory drillholes. In regions where the Ordovician section consists of relatively pure carbonate rocks, as in the southern half of the basin, K-bentonite beds commonly are marked by a prominent deflection on resistivity, gamma-ray, and neutron wireline logs. Cores and drill cuttings are available for a small number of drillholes in the basin. A composite stratigraphic column is shown in Figure 61.

One of the best known Ordovician K-bentonites in the subsurface of the Michigan Basin is in the upper part of the Black

Figure 60. The Black River Group exposed at Bony Falls Dam near Escanaba, Michigan (locality 79). A K-bentonite is exposed in the reentrant near the base of the second chain post from the top of the section.

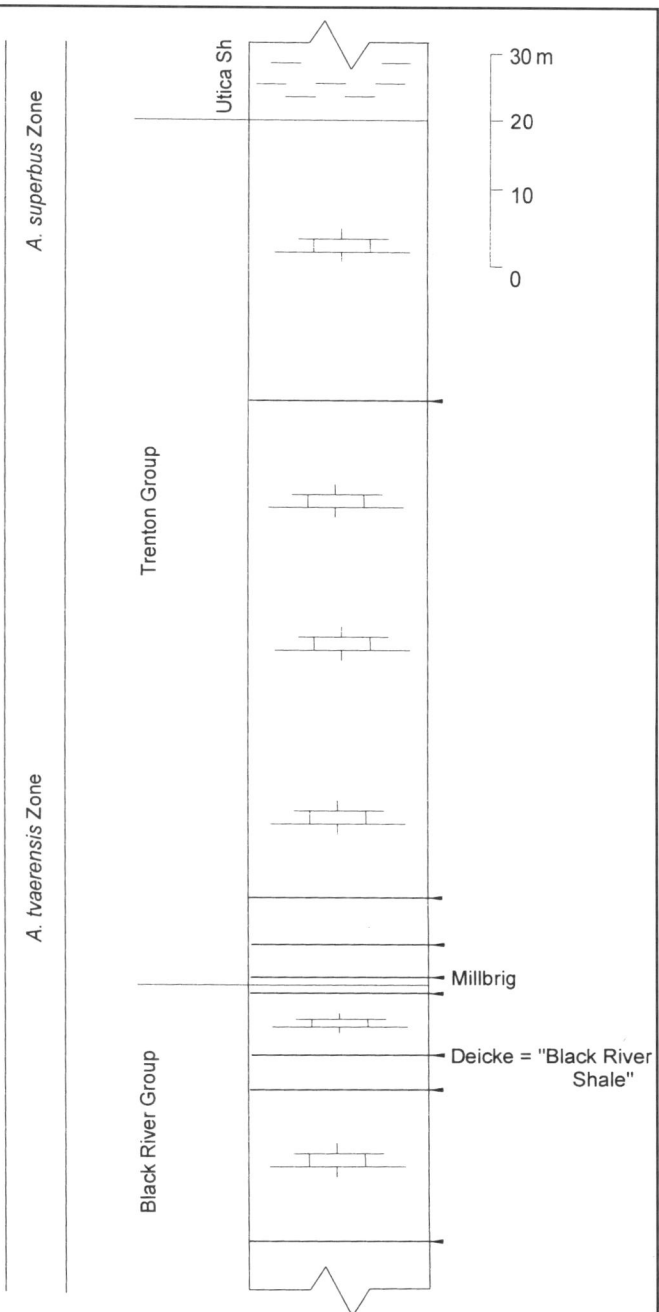

Figure 61. Generalized stratigraphic column of Michigan Basin. Abbreviation: shale, Sh.

River Group and is referred to as the "Black River shale" (Lilienthal, 1978). Typically, the bed occurs 3 m to 10 m below the Trenton–Black River contact. Calvert (1963a, 1963b) correlated the bed on wireline logs, augmented with sample descriptions, from Cass County, Michigan, across northern Ohio to Crawford County, Pennsylvania, and from Jackson County, Michigan, southeastward through Ohio to Wood County, West

Virginia. Likewise, Calvert's (1964) correlation from Fayette County, Ohio, to Brant County, Ontario, shows the "Black River shale" to be widespread and persistent. Chemical analyses of a sample of the "Black River shale" from the ARCO No. 1-14 Dunn drillhole in Calhoun County, Michigan (locality 209), indicate a close chemical affinity with the Deicke K-bentonite. We, therefore, believe the Deicke is equivalent to the "Black River shale." North of approximately 44° North latitude (north of Saginaw Bay) the Black River Group grades to shaly limestone and the Deicke K-bentonite becomes difficult to distinguish on wireline logs.

In addition to the Deicke, several other apparent K-bentonite beds are present near the Black River/Trenton contact in the southern part of the Michigan Basin. These are expressed as prominent deflections on gamma ray/neutron logs as seen, for example, in the Panhandle Eastern Pipeline Company No. 1 Ford Motor Company drillhole in Wayne County, Michigan (Plate 3; locality 139). These K-bentonites are traceable to other wells in the region (Lilienthal, 1978). The K-bentonite succession consists of (1) two sub-Deicke beds in the Black River Group at approximately 35 m and 15 m below the Black River/Trenton contact, (2) a bed situated above the Deicke about 3 m below the top of the Black River, and (3) a prominent bed at or just below the Black River/Trenton contact. The latter K-bentonite has been observed in drill cuttings and cores in the southern part of the Michigan Basin. In the Total Petroleum Inc. No. 2-12 Emil Faist, Jackson County, Michigan (locality 135), this bed occurs at the Black River/Trenton contact, at the 5,231 ft (1,594 m) level, and consists of 20 cm of mixed shale and K-bentonite clay containing biotite. The stratigraphic position of the bed and its biotite content suggest that it is equivalent to the Millbrig K-bentonite. The presence of two or three other K-bentonites in the lower 30 m of the Trenton is suggested by prominent deflections on some wireline logs. In places, it is difficult to distinguish these beds from the associated shaly carbonate rocks.

Within the Trenton Group, approximately 40 m below the contact with the Utica Shale is a persistent deflection on gamma-ray logs (Lilienthal, 1978), believed to be a K-bentonite bed. It is in the same stratigraphic position as the K-bentonite noted by Trevail (1990) in the Sherman Falls Formation in southwestern Ontario.

A thorough biostratigraphic analysis of the Ordovician succession in the Michigan Basin is lacking. One of the very few attempts to zone the subsurface Middle Ordovician was made by Faber (1979) who studied the Viola Holman well, which is ~25 km northwest of the center of Detroit, Wayne County, Michigan. He tentatively placed the *A. tvaerensis–A. superbus* Zone boundary 250 ft (76 m) above the base of the Trenton Limestone at the 3,700 ft level in the core. In a restudy of Faber's samples, Bergström and Mitchell (1992) concluded that the boundary was less precisely known and recommended that it be placed in the interval between 3,702 ft and 3,880 ft in the core. These meager faunal data are consistent with what would be expected by extrapolation from extensively studied stratigraphic sections in nearby Ohio.

St. Lawrence Lowlands and Quebec

Northeast of the Frontenac Arch in the Central St. Lawrence Lowlands of southern Quebec (Fig. 48), K-bentonite beds occur in the Black River and Trenton Groups (Brun and Chagnon, 1979). The oldest known bed is in the upper part of the Leray Formation (Fig. 62) of late *B. compressa* to early *P. undatus* Zone age (Barnes et al., 1981). Near Aylmer, Quebec (locality 82), this bed is as much as 17 cm thick (Fig. 63). It can be correlated eastward to quarries near Joliette (locality 83) and may be equivalent to K-bentonite "B1" described by Bergström and Mitchell (1994) near the top of the Black River Group on wireline logs from the subsurface of southern Quebec. Traced westward in outcrop and on wireline logs, the bed appears to be equivalent to the Deicke K-bentonite ("Black River shale") in southwestern Ontario and in the Michigan Basin.

In the overlying Trenton Group, three K-bentonites have been reported by Brun and Chagnon (1979) from the Montreal Formation and four from the Neuville Formation, primarily in quarries and river sections between Montreal and Quebec City. The lower two K-bentonites in the Montreal are less than 1 cm thick, and the upper bed is 3 cm thick. The lower K-bentonite in the Neuville is less than a centimeter thick and the succeeding beds are 9 cm to 12 cm, 3 cm to 9 cm, and 9 cm to 24 cm thick. The uppermost Neuville K-bentonite is about 30 m below the Utica Shale and is in approximately the same stratigraphic position as the K-bentonite in the Sherman Falls Formation in southwestern Ontario and in the Michigan Basin. The Montreal Formation is believed to be of *C. americanus* Zone age (Barnes et al., 1981), and the Neuville contains *C. americanus* at the base and *O. ruedemanni* in the upper part (Goldman and Mitchell, 1994).

At Neuville, Quebec, northeast of Quebec City, three K-bentonites are present in the Beaupré Formation, a siliciclastic unit lying above the Utica Shale (Belt et al., 1979). The Beaupré contains *C. spiniferus* Zone graptolites. Bergström and Mitchell (1994) proposed bed-for-bed correlations of nine late Mohawkian K-bentonites between Neuville, Quebec, and the Mohawk Valley–Trenton Falls region, New York.

Western Newfoundland

Three K-bentonite beds are present in the early Middle Ordovician Table Head Formation (Fig. 64) on the Port-au-Port Peninsula, western Newfoundland (Finney and Skevington, 1979). The beds are exposed in a 16 m thick section of shaly limestone and calcareous shale in a quarry near West Bay Centre (locality 84). The associated fauna is indicative of the Pacific Province graptolite Zone of *Diplograptus decoratus* (Da3) of the Darriwilian Stage, Llanvirn Series, and also referable to the upper part of the *Paraglossograptus tentaculatus* Zone of the Marathon sequence of west Texas.

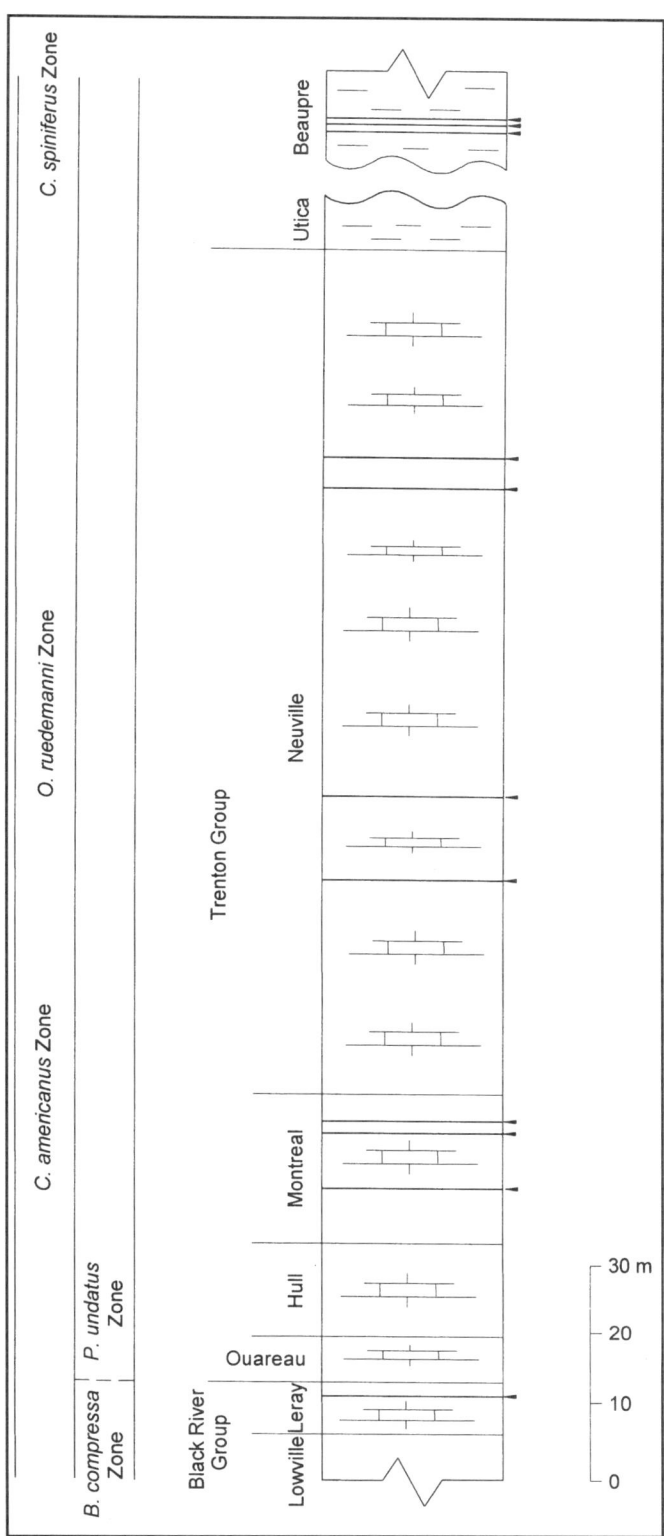

Figure 62. Generalized stratigraphic column of the St. Lawrence Lowlands and southern Quebec.

Figure 63. Gull River Formation exposed in quarry at Aylmer, Quebec (locality 82). Warren D. Huff points to a 17-cm-thick bed believed to be the Deicke K-bentonite.

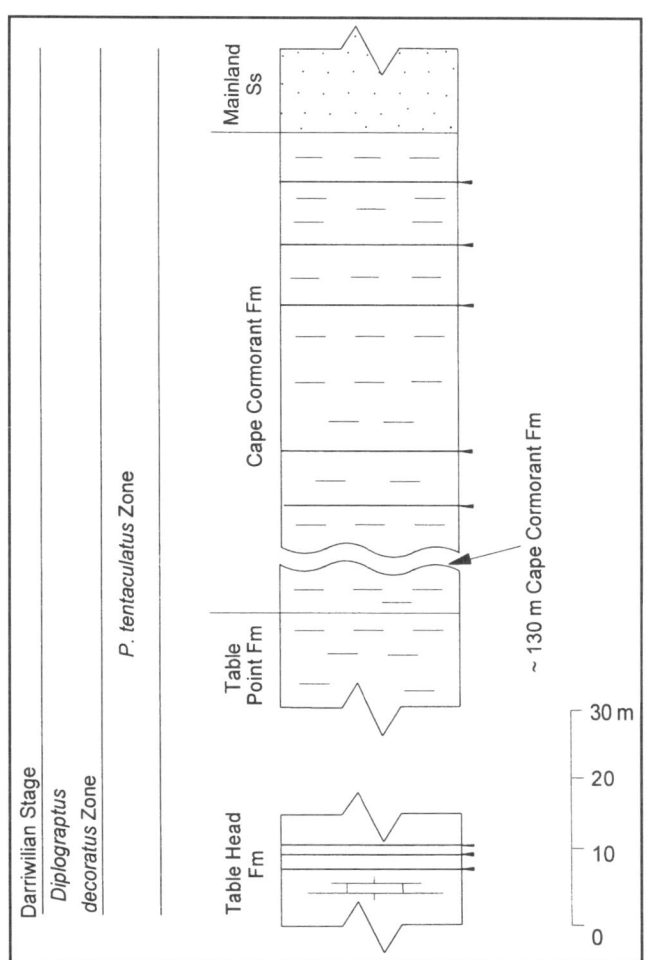

Figure 64. Generalized stratigraphic column of the Port-au-Port Peninsula, western Newfoundland. Abbreviations: Formation, Fm; and sandstone, Ss.

At least five K-bentonites, less than 5 cm thick, are present in the somewhat younger Cape Cormorant Formation (Fig. 64) on the Port-au-Port Peninsula (locality 85). The Cape Cormorant is believed to be late Llanvirn (pre-*H. teretiusculus*, probably Da4) in age (Charles E. Mitchell, 1994, personal communication).

Eastern Oklahoma–Arkoma Basin

In several drill holes in the Arkoma Basin of eastern Oklahoma, gamma-ray/neutron logs show prominent deflections at the top and bottom of the Corbin Ranch "submember" of the Pooleville Limestone Member of the Bromide Formation (Fig. 65) that probably correspond to K-bentonite beds. These deflections are particularly noticeable locally in the subsurface of Seminole and Pottawatomie Counties (localities 210 and 211). These beds may be exposed at the surface in the Arbuckle Mountains of southern Oklahoma. Recent analysis of a clay bed at the Viola-Bromide contact, described in outcrops (localities 86 and 87) by Decker (1933), show the bed to consist of mixed layer illite/smectite clay, indicative of a K-bentonite (Christopher L. Johnson, 1994, personal communication). Wireline signatures bear a striking resemblance to the K-bentonite–bearing interval seen at the contact between the Black River and Trenton Groups over a wide area of the eastern Midcontinent (Fig. 66). Graphic correlation data based on conodonts suggest that the Corbin Ranch "submember" is latest Turinian to mid-Chatfieldian in age (Amsden and Sweet, 1983) and therefore is correlative with strata in the interval near the contact between the Black River and Trenton Groups. In the Corbin Ranch, *P. undatus*, a characteristic component of the late Middle Ordovician conodont fauna in the eastern Midcontinent, is abundant.

The Viola Springs is separated from the underlying Bromide Formation by a sharp lithologic contact that shows evidence of pre-Viola induration and dissolution (Amsden and Sweet, 1983). There is also biostratigraphic evidence for a hiatus at this horizon. The localized occurrence of K-bentonite beds in outcrop and in the subsurface suggests that volcanic ash deposits and enclosing Corbin Ranch strata were cut out by pre-Viola erosion in some areas of eastern Oklahoma. In Arkansas, K-bentonites are also absent from this stratigraphic interval in the outcrop belts. The disconformity between the Bromide and Viola is very likely the same one that marks the contact between the Black River and Trenton Groups elsewhere in the Midcontinent. Stratigraphic relations suggest that the K-bentonites in the Corbin Ranch are equivalent to the widespread Deicke and Millbrig K-bentonite Beds. Overall wireline log character of the Middle and Late Ordovician stratigraphic succession in eastern Oklahoma is remarkably similar to that seen in the eastern Midcontinent region. However, we did not chemically fingerprint the Oklahoma K-bentonites.

Western Iowa, eastern Kansas, and Nebraska

Ordovician rocks are present in the subsurface of western Iowa and the eastern parts of Kansas and Nebraska, but little is known about the presence of K-bentonite beds. We suspect that some beds may be present locally in this region because they occur in Ordovician strata nearby in parts of Iowa, Missouri, and Oklahoma. Of particular interest, is the Middle Ordovician succession including the Platteville, Decorah, and Viola strata (Fig. 67).

The Deicke K-bentonite has been identified by chemical fingerprinting techniques in the subsurface of western Iowa (locality 213; Kolata et al., 1986, 1987). The bed is about 5 cm thick and occurs in the upper part of the Platteville Formation (Fig. 68) in a stratigraphic succession that is similar to that observed in the eastern part of Iowa (Witzke, 1980). This is the northwesternmost known occurrence in North America of an unequivocal Ordovician K-bentonite with a volcanic source in the Iapetus region.

A possible K-bentonite was described in subsurface samples from a well in Wabaunsee County, Kansas, by Taylor (1947). Drill cuttings from the basal part of the Viola Limestone in the Empire Oil Company No. 1 Schwalm well (locality 212)

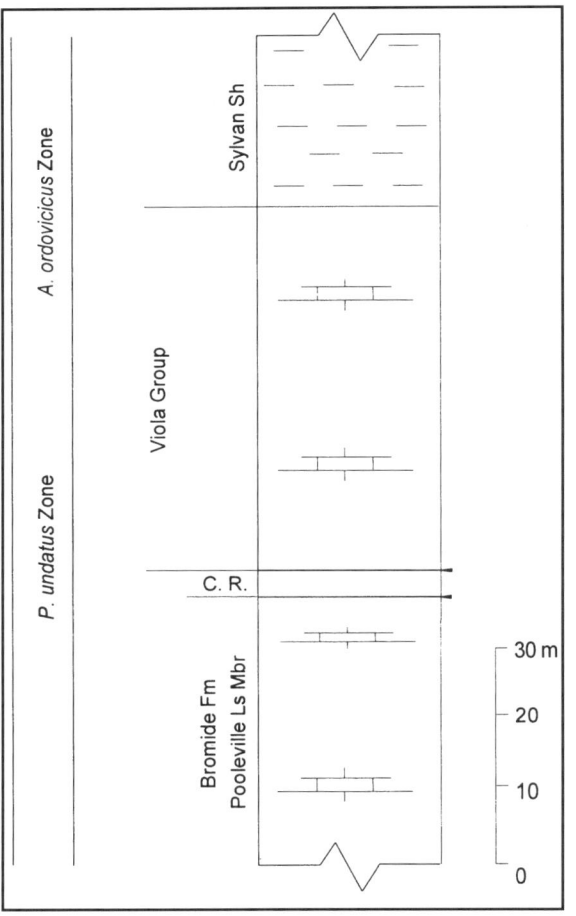

Figure 65. Generalized stratigraphic column of the Arkoma Basin in eastern Oklahoma. Abbreviations: Corbin Ranch "submember," C.R.; Formation, Fm; and sandstone, Ss.

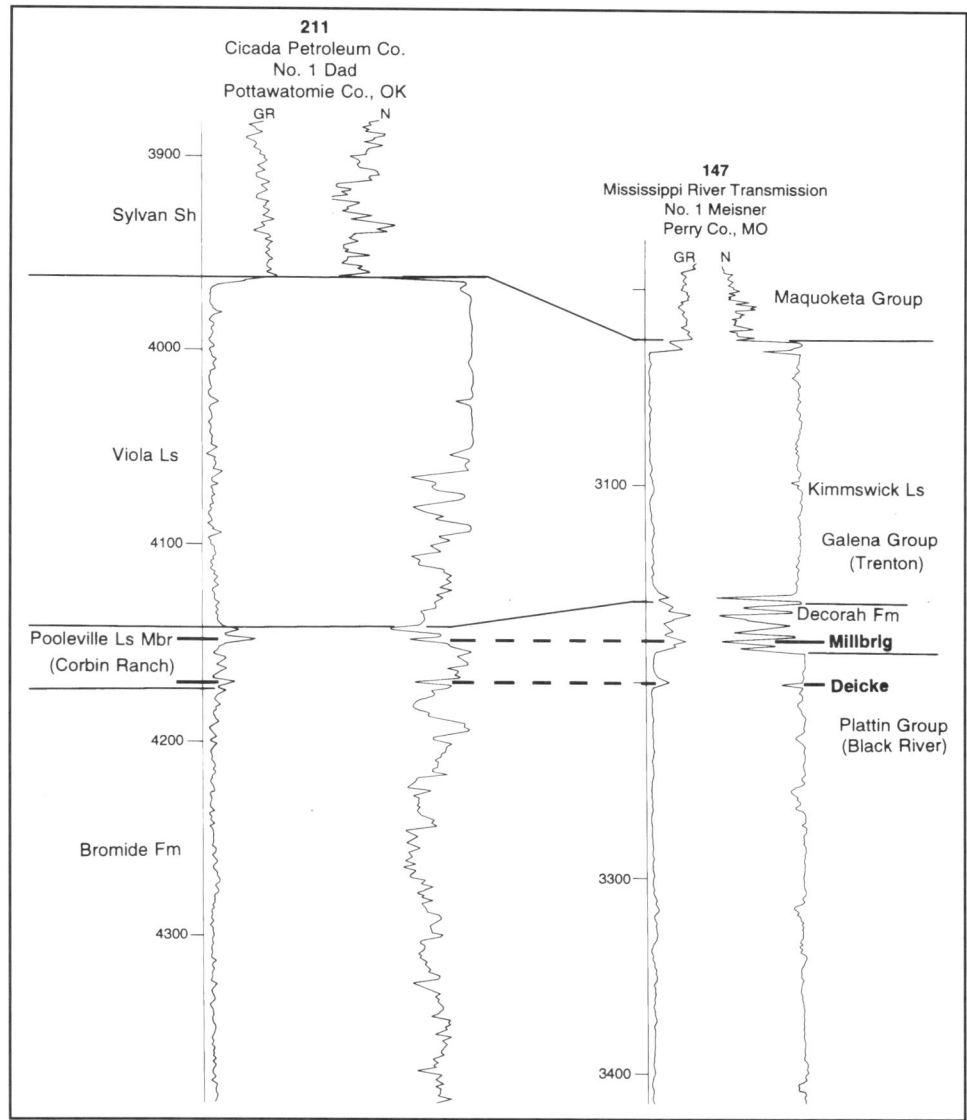

Figure 66. Gamma-ray (GR)/neutron (N) log correlation between Pottawatomie County, Oklahoma, and Perry County, Missouri, showing suspected correlation of the Deicke and Millbrig with K-bentonites in the Corbin Ranch beds of the Pooleville Limestone Member of the Bromide Formation. Depth in feet. Abbreviations: Formation, Fm; limestone, Ls; and shale, Sh.

contain "gray bentonite-like shale" apparently in two beds separated by about 3 m of limestone. Based strictly on lithologic similarity, Templeton and Willman (1963) believed these K-bentonite–bearing strata were equivalent to the Buckhorn and St. James Limestone Members of the Dunleith Formation in the Upper Mississippi Valley region. Green and brown waxy shale were also observed in cuttings from wells nearby in Saline and McPherson Counties, Kansas, at the same stratigraphic position (Taylor, 1947, p. 1268). Taylor stated that "fragments from some of these shaly layers will, if dropped in water, swell and crumble in the manner of bentonites." He noted that altered volcanic ash was known from the eastern United States, implying that they may extend into Kansas. We suspect that if K-bentonites are present they are <10 cm thick.

Unfortunately, the Middle Ordovician succession of Kansas and Nebraska contains a large amount of shale, particularly in the Platteville and Decorah intervals, making it difficult to distinguish K-bentonite beds on wireline logs.

Texas

Available evidence suggests that volcanic rock fragments and possibly K-bentonites may be present in outcrops and in the subsurface Ordovician succession of Texas. A notable ref-

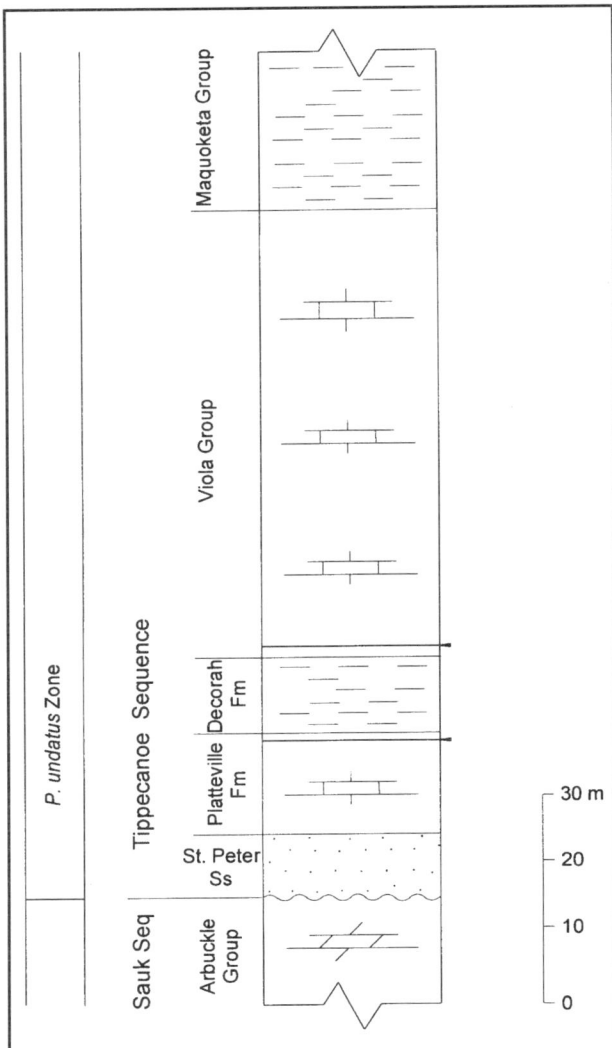

Figure 67. Generalized stratigraphic column of western Iowa and eastern Kansas and Nebraska. Abbreviations: Formation, Fm; Sequence, Seq; and sandstone, Ss.

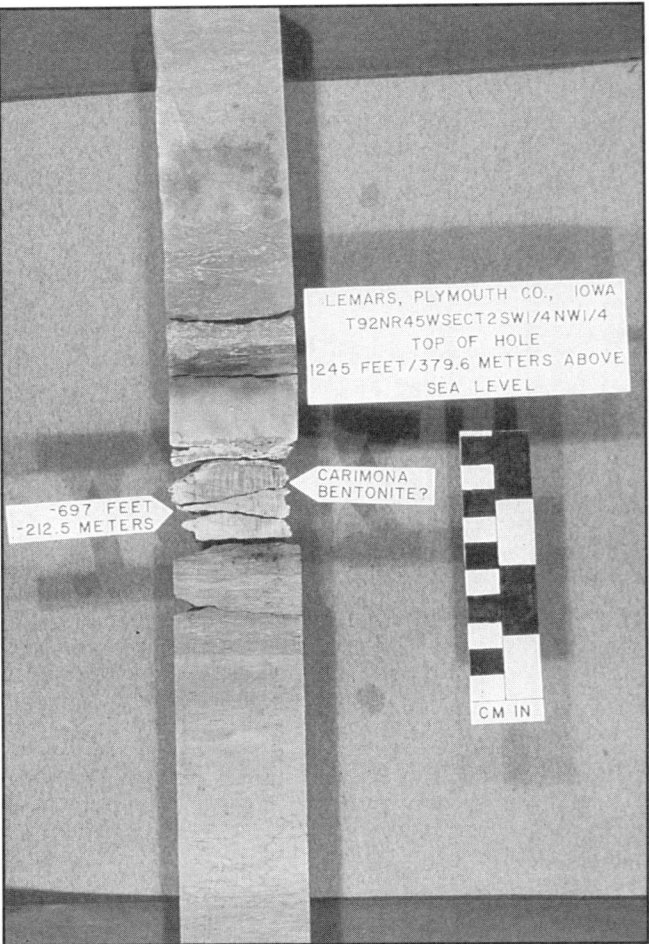

Figure 68. Drill core from the upper part of the Platteville Group at Le Mars, northwestern Iowa (locality 213). The Deicke K-bentonite (labeled Carimona bentonite?) is a 4-cm-thick bed of compacted clay. Photograph courtesy of Steve Schutter.

erence to Ordovician K-bentonites was made by King (1937) in the Marathon region of southwestern Texas. King reported three "light-green hard bentonite" beds in the Lower Ordovician Marathon Limestone in the King & Franklin Gage No. 1 well in Brewster County. If these in fact are K-bentonites, they would be the oldest Ordovician K-bentonite beds known in North America. The Marathon Limestone also contains local concentrations of biotite, plagioclase, mixed-layer clay, arkose, and trachyte boulders interpreted by Young (1970) to be evidence of Early Ordovician volcanic activity within or near the Marathon region. The presence of andesite and dacite rock fragments and bentonite-derived clays in Lower and Middle Ordovician rocks of the Texas/Mexico border area may be further evidence of volcanism near the Marathon region (Dickerson, 1994).

REGIONAL CORRELATION OF K-BENTONITES

The stratigraphic distribution of Ordovician K-bentonites in eastern North America is summarized in Plate 1. The age and relative position of beds is established primarily on the basis of associated Midcontinent or North Atlantic conodonts and/or North American or European graptolites. Correlation of individual beds was accomplished by a variety of methods including chemical fingerprinting, matching detailed outcrop descriptions, tracing on wireline logs, and studying drill cores and drill cuttings. Most of the 23 columns represent a composite section for a given region. In general, the number and thickness of beds and grain size of the Middle Ordovician K-bentonites all decrease north and west of the southern and central Appalachians, suggesting that the source volcanoes were situated east of North America at a latitude between Alabama and North Carolina. It is possible, however, that the late Mohawkian and early Cincinnatian K-bentonites in up-state New York and Quebec were derived from volcanoes situated at a latitude between eastern Pennsylvania and maritime Canada. The following discussion of regional correlations is organized from the oldest to youngest K-bentonite occurrences.

Correlation of the Mohawkian Series, which contains the largest concentration and most widespread Ordovician K-bentonites in eastern North America, is established in a series of wireline cross sections (Plates 2 through 7) that tie key outcrop sections to significant deep drillholes. Many more wireline logs than are shown in this book were used to establish the correlations.

Ibexian Series

The oldest reported K-bentonites in the Ordovician of North America are in the Ibexian Marathon Limestone in the Marathon region of south central Texas (King, 1937). The three beds reported from a single drill hole have not been analyzed to confirm that they are altered volcanic ash. During the final preparation of this book, a 2 cm thick K-bentonite was discovered in the upper part of the Marathon Limestone in the Solitario region of southern Texas (Patricia W. Dickerson, 1996, personal communication).

A notable K-bentonite in the upper part of the Knox Group can be traced in the subsurface of eastern Missouri through the southern Illinois Basin into north-central Kentucky where it appears to correlate with one of two K-bentonites in the Beekmantown Formation. Given the lack of biostratigraphic control and lack of sufficient subsurface samples to chemically correlate the beds, we cannot determine which of the two beds might extend into Missouri. For the sake of illustration, we show the Missouri and Illinois Basin bed aligned with the upper of the two beds in Kentucky, but it just as likely correlates with the lower bed (Plate 1). Both beds produce a prominent deflection on gamma-ray/neutron logs and at some localities drill cuttings contain biotite flakes. The two K-bentonites reported from the Knox Group in the Black Warrior Basin (Thomas, 1972, 1988) are the same age as those in Missouri, Illinois, and Kentucky and are shown to be equivalent on Plate 1. They occur in the *Oepikodus evae* North Atlantic Province. We suspect that these beds are much more widespread and may be present in outcrop and/or the subsurface in the southern Appalachians and, possibly, the Marathon region of west Texas.

Whiterockian Series

A small number of Whiterockian K-bentonites are known from eastern North America. A major difficulty in correlating these beds results from the diachronous age of the sub-Tippecanoe unconformity and the equally diachronous nature of the overstepping Tippecanoe sequence. For example, in the upper Mississippi Valley region, the early Mohawkian St. Peter Sandstone, with a K-bentonite at its base, overlies the Ibexian Shakopee Dolomite (Fig. 11). A major hiatus, spanning most of Whiterockian time, separates the two formations. Consequently, the precise age of the basal St. Peter K-bentonite is unknown and its correlation with Whiterockian K-bentonites elsewhere in eastern North America is uncertain. The age of the K-bentonite that unconformably overlies the Beekmantown Formation in eastern Tennessee (Fig. 39) is somewhat better constrained. Within the region, the Lenoir Limestone containing conodonts of the *Pygodus serra* Zone overlies the K-bentonite fixing the age as early to middle Whiterockian. A K-bentonite of approximately the same age is present in the lower part of the Athens Shale east of the Helena fault near Calera, in northeastern Alabama (locality 54). The Calera bed occurs in strata bearing conodonts characteristic of the lower *E. reclinatus* Subzone, *P. serra* Zone (Schmidt, 1982). The base of the Athens Shale in this region corresponds to a level in the lower *Hustedograptus teretiusculus* graptolite Zone (Finney, 1983). The K-bentonite in the Athens and Lenoir may correspond to a locally occurring bed in the Wells Creek Dolomite of northern Kentucky.

At Mosheim, Tennessee, five K-bentonites are present in the lower part of the Blockhouse Shale (Fig. 39) with an age corresponding to early *E. robustus* Subzone, *P. serra* Zone (Fetzer, 1973; Bergström, 1973; Kurapkat, 1986). The eight K-bentonites in the Table Head and Cape Cormorant Formations in western Newfoundland (Fig. 64) are also late Llanvirn in age, but the Newfoundland K-bentonites appear to be slightly older based on the presence of graptolites referable to the *P. tentaculatus* Zone.

The K-bentonite bed in the Little Oak Limestone at Ragland, Alabama, and the single bed in the upper part of the Blockhouse Shale at Mosheim, Tennessee, are associated with the *P. anserinus* Zone fauna. We tentatively correlate the two occurrences.

Between 20 and 25 K-bentonites with a cumulative thickness of 7 m to 8 m have recently been discovered in the Fort Peña Formation (*P. tentaculatus* Zone) in the northern part of the Solitario region of southern Texas (Patricia W. Dickerson, 1996, personal communication).

Mohawkian Series

The Mohawkian Series contains the most widespread and persistent K-bentonites in eastern North America. K-bentonite R1 observed by Miller and Fuller (1954) in the Dot Limestone of Lee County, Virginia (Fig. 37), is early Mohawkian based on the presence of *P. aculeata* (Hall, 1986). The bed also is believed to be present in the Dot Limestone in the subsurface of Lee County where it occurs 8 m above the base (Harris, 1967). A biotite-rich K-bentonite in the upper part of the Blackford Formation in the central outcrop belt near Blackford, Russell County, Virginia (Heyman, 1970), probably is equivalent to R1. Bed S1 (Fig. 44), described by Perry (1964), occurring just above the contact between the Beekmantown and overlying St. Paul Group in Pendleton County, West Virginia, may be equivalent to R1 but this is somewhat uncertain because of sparse faunal control. Miller and Fuller (1954) suggested that R1 is equivalent to bed B-1 described by Fox and Grant (1944) from the Murfreesboro Limestone near Chattanooga, Tennessee. This correlation seems reasonable because the Murfreesboro in nearby Kentucky also belongs to the *P. aculeata* Chronozone (Sweet, 1984). Furthermore, the biotite-rich K-bentonite in the Dutchtown Limestone of southeastern Missouri is late Whiterockian in age and possibly equivalent to R1. We show the K-bentonite at the base of the St. Peter (Plate 1) in the upper Mississippi Valley to be equivalent to R1, but this is highly speculative because of the poor time constraints on the St. Peter bed. We believe the R1 and its equivalents have potential for regional correlation.

One or another of the seven K-bentonites of *B. gerdae* Subzone age that occur in the Botetourt and Edinburg Formations of northern Virginia may correspond to the single bed occurrences in the Hurricane Bridge Limestone of southwestern Virginia (bed R2 of Miller and Fuller, 1954), Athens Shale of northeastern Alabama, and Beckett Limestone of Missouri.

As many as 60 middle to late Mohawkian K-bentonites, all shown on Plate 1, are known from eastern North America, most of which are thin, locally occurring beds that can not be traced with confidence for great distances. The beds with the greatest event-stratigraphic potential include the well-known Deicke and Millbrig K-bentonites. They provide a stratigraphic reference for the identification and discussion of other significant Mohawkian K-bentonites.

The lateral continuity of the Deicke and Millbrig is demonstrated on a network of cross sections compiled primarily from wireline logs (Plates 2 through 7). Both beds produce prominent and consistently recognizable deflections on the logs. The regional correlation of both beds was demonstrated by Huff and Kolata (1990) on a similar but less extensive network of cross sections. Likewise, although Calvert (1962, 1963a, 1963b, 1964) did not specifically identify the Deicke and Millbrig, it is clear from his wireline cross sections and accompanying drill-cutting descriptions that he was able to correlate these beds in the subsurface over a wide region of eastern North America. Both K-bentonites lie in the *P. undatus* Chronozone (Sweet, 1984). In terms of the conodont-based graphic correlation scheme (Sweet, 1984), the Deicke projects into the composite standard section (CSS) at a level of 971 m and the Millbrig at 975 m (Walter C. Sweet and Thomas H. Shaw, 1994, personal communications).

Deicke K-bentonite Bed. The Deicke can be traced confidently from its type section near St. Louis, Missouri (Willman and Kolata, 1978), into the subsurface approximately 30 km to the northeast where it produces a prominent deflection on the gamma ray/neutron log from the Laclede Gas Company No. 1 Mintert well in St. Charles County, Missouri (locality 105). This is a key well in correlating the Deicke throughout the rest of eastern North America. The bed typically is 5 cm to 10 cm thick and occurs between 1.5 m and 2 m below the top of the Platteville Group at outcrops in the St. Louis area. North of St. Louis the Deicke can be identified on wireline logs from Pike County, Missouri, northward through western Illinois to the New Jersey Zinc Company No. H-1 Cochran well in Hancock County, Illinois (locality 112), where the bed was chemically identified from a drill-core sample (Kolata et al., 1986). Similarly, the Deicke extends through the subsurface of eastern Iowa to the drill core from the Big Springs BS-2 well in Clayton County (locality 116). The stratigraphic succession comprising the Platteville and Galena Groups in this well is essentially the same as that observed in the nearby outcrop belt of northeastern Iowa, southwestern Wisconsin, and southeastern Minnesota. This is significant because it demonstrates that the Deicke K-bentonite and associated strata can be correlated from eastern Missouri through the subsurface into the outcrop belt of the Upper Mississippi Valley. Available gamma-ray/neutron logs from north-central Iowa and south-central Minnesota do not show a prominent deflection, perhaps because the Deicke is relatively thin. The bed is present in a core from the Northern Natural Gas Company No. 67-1A Hollandale well in Freeborn County, Minnesota (locality 118), but is not well defined on the gamma-ray wireline log from this drill hole (Plate 2).

In Plymouth County, northwestern Iowa (locality 213), the Deicke has been identified by chemical fingerprinting methods in a drill core in the upper part of the Platteville Group (Kolata et al., 1986). Even though this is one of the most distal occurrences of the Deicke, at 5 cm thick, the bed is comparable in thickness to most localities in the Mississippi Valley. Given the thickness and wide distribution, it is conceivable that the Deicke extends westward and is yet to be discovered in the Winnepeg Formation in the Williston Basin of North Dakota (Fig. 1). Furthermore, future investigations may show that the Deicke is equivalent to one of the two beds described herein from the upper part of the Bromide Formation in the Arkoma Basin of southern Oklahoma and in the Platteville, Decorah, and Viola of eastern Nebraska and Kansas (Fig. 67).

Forming a prominent deflection on gamma-ray logs, the Deicke is readily traceable from its type area of eastern Missouri northeastward through the subsurface of the Illinois Basin. The Deicke pinches-out on the Kankakee Arch in northern Illinois and northern Indiana, but it is present north of the arch in the Michigan Basin where it produces a prominent deflection on gamma ray/neutron logs (Plate 3). In southwestern Ontario, the Deicke is present in the upper part of the Coboconk Formation (Fig. 56) of the Black River Group where it appears to correlate with K-bentonite MH of Liberty (1969). The Deicke diminishes in thickness eastward through southern Ontario. As noted by Bergström and Mitchell (1994), the Deicke may be equivalent to their K-bentonite "B1" in the Leray Formation north of Montreal and possibly to a K-bentonite in the Chaumont Formation of New York (Kay, 1935).

The Deicke is readily traceable on wireline logs in the subsurface of the Illinois Basin of southern Illinois, southern Indiana, and western Kentucky eastward to the Cincinnati Arch in southern Ohio (Plate 4) where it is equivalent to "marker bed β"

(Stith, 1979). Wireline logs (Plates 4, 5, 7) show the Deicke to be equivalent to the "Pencil Cave bentonite" in the upper part of the Tyrone Limestone of Kentucky, T-3 (Wilson, 1949) in the upper Carters Limestone in Tennessee and Alabama, and R7 in the Eggleston Formation of western Virginia (Huff and Kolata, 1990). The Millbrig also has been shown to correlate with V-3 (Rosenkrans, 1936) in the Moccasin Formation in the central outcrop belt of western Virginia (Haynes, 1992, 1994). These correlations are supported by studies of phenocryst mineralogy in outcrops and cores along the Cincinnati Arch and in the southern Appalachians (Haynes, 1994). Due to abrupt facies and thickness changes in the Middle Ordovician succession of eastern Ohio, it is difficult to confidently correlate the Deicke and other K-bentonites on wireline logs eastward into Pennsylvania or West Virginia. Likewise, even though the Deicke and associated K-bentonites are easily identified and correlated in the Michigan Basin and southern Ontario, it is not readily apparent how these beds correlate through the subsurface of central and northern Ohio. It is a challenge to correlate individual beds north of Delaware County, Ohio (confirmed by David A. Stith, 1994, personal communication).

Millbrig K-bentonite Bed. The Millbrig K-bentonite can be correlated from its type section near Galena, Jo Daviess County, Illinois (Willman and Kolata, 1978), throughout the Upper Mississippi Valley outcrop belt (Kolata et al., 1986, 1987) into the subsurface to the Big Springs BS-2 well in Clayton County, Iowa (Plate 2). Locally, in the subsurface of the Mississippi Valley, the Millbrig is difficult to identify on wireline logs because the confining Spechts Ferry Shale Member of the Decorah Formation generally produces an equally high gamma-ray response that obscures the K-bentonite. Where observed in cores, however, the Millbrig is present near the middle of the shale.

In the outcrop belt of eastern Missouri, the Millbrig is a persistent bed, locally as much as 15 cm thick, in the Spechts Ferry Shale Member of the Decorah. The Millbrig is not present in Grays Point Quarry (locality 40) in Scott County, Missouri, nor nearby in the subsurface in the Shell Oil Co. No. 1 Trail of Tears State Park well in Cape Girardeau County (locality 148).

The Spechts Ferry Shale pinches out along the western margin of the Illinois Basin. Consequently, in the basin, the pronounced lithologic contrast between the Millbrig and confining limestones results in a prominent deflection on wireline logs that is readily traceable from eastern Missouri northeastward through the Illinois Basin (Plate 3). The bed produces a particularly high gamma-ray response on logs in Piatt and Champaign Counties, Illinois. Like the Deicke K-bentonite, the Millbrig pinches-out in the region of the Kankakee Arch in northeastern Illinois and northwestern Indiana. Late Turinian to early Chatfieldian uplift and erosion of the arch prevented preservation of the ash beds. On the north side of the arch in the Michigan Basin, the Millbrig is present in the basal Trenton succession in Jackson County, Michigan, and at some localities farther to the east in the basin center; however, due to the shaliness of the Trenton carbonate rocks in the Michigan Basin, the Millbrig is locally difficult to distinguish on wireline logs. This is a particularly difficult problem in the northern part of the basin where the section grades to mostly shale.

The Millbrig is present in the basal part of the Kirkfield Formation of the Trenton Group in southwestern Ontario (Trevail, 1990), but becomes increasingly more difficult to recognize east of Toronto. The Millbrig appears to correlate with K-bentonite MR of Liberty (1969). Both the Deicke and Millbrig have been identified by geochemical fingerprinting methods in the Ontario Geological Survey 82-2 drill core in Kent County, Ontario (locality 207; Trevail, 1990).

The Millbrig can be traced on wireline logs from eastern Missouri across the southern Illinois Basin into southern Ohio and northern Kentucky (Plates 4 and 5) where it is equivalent to "marker bed α" of Stith (1979). In the region of the Cincinnati Arch in north-central Kentucky, the Millbrig is locally absent, apparently resulting from pre-Lexington erosion (Cressman, 1973). Similar stratigraphic relations occur in central Tennessee. For example, near South Carthage in Smith County (locality 48) the Millbrig is present at the base of the Nashville Group (=Trenton). Southeast of here in the Sequatchie Valley, however, the Millbrig is overlain by as much as 2 m of Carters Limestone (=Black River). Furthermore, the Millbrig and, locally, the Deicke are absent near the crest of the Nashville Dome in Giles, Maury, and Marshall Counties as a result of pre-Nashville (pre-Trenton) erosion (Wilson, 1949). This erosional unconformity apparently passes laterally into a conformable contact between the Stones River and Nashville Groups in much of the western Valley and Ridge of Tennessee, Georgia, and Alabama (Haynes, 1994). Likewise, Haynes noted that the unconformity in central Kentucky passes eastward into a conformable surface within the Eggleston and Trenton succession in the Valley and Ridge of southwestern Virginia and northeastern Tennessee. Farther east the unconformity merges with a conformable surface near the contact between the Bays and Trenton Formations (Kreisa, 1980).

As shown by Haynes (1992, 1994), the Millbrig correlates with R10 (Miller and Fuller, 1954; Miller and Brosgé, 1954) in the Rose Hill district of southwestern Virginia and with V-4 (Rosenkrans, 1936) in the adjacent thrust sheet to the east, also referred to as the central outcrop belt of the Valley and Ridge Province.

In the northeastern Kentucky counties of Greenup, Rowan, Bath, and Montgomery, the Millbrig is either absent or very thin and does not produce a well-defined deflection on gamma-ray logs (Plate 7). However, the bed is well developed on logs from Lincoln County, Kentucky, to Jackson County, Alabama, and is traceable in outcrop southward to Birmingham, Alabama (locality 53), where it is present in the Chickamauga Limestone of the Stones River Group. Haynes (1994) correlated the Millbrig with a biotite-rich bed in the Colvin Mountain Sandstone of the Blount detrital wedge in northeastern Alabama and northwestern Georgia. Wireline logs in Floyd and Pike Counties, Kentucky, and Dickenson and Russell Counties, Virginia (Plate 5), show an apparent K-bentonite bed between 3 m to 5.5 m below the Millbrig. This bed is not present, however, in outcrop at Hagan, Lee County, Virginia, nor in the nearby Shell Oil Co. No. 1 L. S. Bales well (localities 59 and 164). However, Miller and Fuller (1954) describe two K-bentonites (R8 and R9) from this interval at other localities in Lee County, Virginia. One or two relatively thin K-bentonites are also known to occur between the Deicke and Millbrig in parts of central Tennessee.

We were not able to confidently correlate the Deicke or Millbrig on wireline logs in the subsurface of West Virginia, Pennsylvania, and New York. Furthermore, although K-bentonite beds are known from the upper Middle Ordovician outcrops in these states, correlations are inconclusive. The uncertainties of correlating K-bentonites through the subsurface from West Virginia to upstate New York were also noted by Wagner (1961). If the Deicke and Millbrig are present in this region, they are thinner than farther south in the southern Appalachians or westward toward the Cincinnati Arch. Presumably, both the Deicke and Millbrig ashes thinned abruptly northward from southwestern Virginia away from the main downwind plume that probably originated in the region of eastern Georgia or South Carolina.

Recently, Ryder (1991, 1992a, 1992b) and Ryder et al. (1992) compiled four regional diagrammatic cross sections showing the stratigraphic framework of the Cambrian and Ordovician Systems in the central Appalachian basin. The sections are based on significant deep drill holes in parts of Ohio, Pennsylvania, West Virginia, and Virginia. Several K-bentonites are correlated through the region, including Stith's (1979, 1986) α and β markers (our Millbrig and Deicke). The α marker is correlated with K-bentonite S-2 (Perry, 1964) in the Nealmont Limestone of Rockingham County, Virginia, and with K-bentonite R (Thompson, 1963) in the Coburn Formation of Clinton County, Pennsylvania (Ryder, 1992b; Ryder et al., 1992). Also, marker β is correlated with Rosenkrans's (1934a) bed No. 0 in the Nealmont Limestone of central Pennsylvania (Ryder, 1992b). The central Pennsylvania correlations are in conflict with McVey (1993) who showed by chemical fingerprinting methods on samples from a series of outcrops from southwestern Virginia to central Pennsylvania that the Deicke and Millbrig are equivalent to beds No. 2 and 4 in the lower part of the Salona Formation. Ryder's correlations could not be confirmed nor compared to our work because the wireline logs were not published with his cross sections.

Other widespread Mohawkian K-bentonite beds. In the outcrop belt of eastern Missouri, at least four K-bentonite beds are present locally in the Turinian Platteville/Plattin Groups below the Deicke (Fig. 18). Some wireline logs from deep wells penetrating the Platteville Group nearby in the Illinois Basin locally show thin sharp deflections as much as 40 m below the Deicke K-bentonite. These deflections may correspond to the Missouri K-bentonites, but they are not as persistent as the Deicke or Millbrig (Plates 3, 4, 5). Stith (1979, 1986) observed three K-bentonites, designated in ascending order "b," "a," and "γ," in the same stratigraphic interval in southern Ohio.

These beds can be traced eastward through the subsurface into the southern Appalachians where they are somewhat thicker and more persistent.

Ocoonita K-bentonite Bed. In southern Ohio, "marker bed γ" (Stith, 1979) occurs approximately 6 m below the Deicke and can be traced on wireline logs southeastward from Highland County, Ohio, through eastern Kentucky into western Virginia (Plates 5, 6). The interval of limestone between the γ K-bentonite and the Deicke gradually thickens from 6 m in Highland County, Ohio, to 12 m in Lee County, Virginia. Judging from outcrop and subsurface data, K-bentonite γ is the same bed recognized by Miller and Fuller (1954) and Miller and Brosgé (1954) as K-bentonite R6 in outcrops in the Rose Hill district of Lee County, Virginia. At Hagan, Virginia (locality 59), R6 is 15 cm thick and occurs in the Eggleston Limestone 12 m below the Deicke. Approximately 10 km east of Hagan in the Shell Oil Co. No. 1, L. S. Bales well (locality 164), a prominent deflection on the gamma ray log at 12 m below the Deicke corresponds to R6. In recognition of the regional distribution of this bed, it is herein formally named the "Ocoonita K-bentonite Bed" after the village of Ocoonita, Lee County, Virginia, situated 14 km northeast of the type section at Hagan, Virginia.

Rosenkrans's (1936) V-2 K-bentonite in the Moccasin Formation in the central outcrop belt of southwestern Virginia is in the same stratigraphic position and possibly equivalent to the Ocoonita K-bentonite. Furthermore, a detailed comparison of wireline logs in Ohio, Kentucky, and Tennessee suggests that the K-bentonite described by Wilson (1949) as T-2 in the Carters Limestone of Tennessee, at an average of 6 m below the Deicke (T-3), is probably equivalent to the Ocoonita K-bentonite Bed. In addition, K-bentonite B-2 of Fox and Grant (1944) in the Carters Limestone near Chattanooga, southeastern Tennessee, is equivalent to the Ocoonita.

The Ocoonita K-bentonite can be traced on wireline logs westward into the Illinois Basin (Plates 4 and 5) and northward from Illinois into the Michigan Basin (Plate 3). The bed is particularly well developed in Douglas, Piatt, and Champaign Counties, Illinois. A possible equivalent bed is present 4 m below the Deicke in the Zell Limestone Member of the Macy Limestone in eastern Missouri (Fig. 18). In the Michigan Basin, a bed in the same stratigraphic position produces a prominent deflection on the logs from Eaton, Sanilac, and Wayne Counties, Michigan, and Essex and Elgin Counties, Ontario. In the Middle Ordovician outcrop belt of southern Ontario, the Ocoonita appears to correlate with Liberty's (1969) MX K-bentonite in the upper part of the Gull River Formation.

Hockett K-bentonite Bed. The K-bentonite identified by Stith (1979, 1986) as "marker bed b" also can be traced confidently on wireline logs from southern Ohio through eastern Kentucky to western Virginia (Plates 5 and 6). In southern Ohio, the bed typically occurs about 23 m below the Deicke K-bentonite. The interval remains approximately the same in eastern Kentucky, thickening slightly in western Virginia. Marker bed b correlates with K-bentonite R5 (Miller and Fuller,

1954; Miller and Brosgé, 1954) in the Rose Hill district of Lee County, Virginia. In recognition of its wide distribution, the bed is herein formally named the "Hockett K-bentonite Bed" for the village of Hockett, Virginia, situated 9 km northeast of the type section at the Hagan railroad siding (locality 59). At the type section, the Hockett K-bentonite is a 15 cm thick bed in the upper part of the Hardy Creek Limestone, 30 m below the Deicke K-bentonite. In the nearby Shell Oil Co. No. 1 L. S. Bales well, the Hockett is marked by a prominent deflection on the gamma-ray log at 33 m below the Deicke. The Rockvale "metabentonite" (Conkin and Conkin, 1992), occurring in the upper part of the Lebanon Limestone at 24 m below the Deicke at Lebanon, Williams County, Tennessee, may correlate with the Hockett K-bentonite, but this has yet to be established with certainty. The Hockett K-bentonite is present locally in the Illinois Basin (Plate 4) and is in approximately the same stratigraphic position as an unnamed K-bentonite at the top of the Mifflin Formation (Willman and Kolata, 1978) in the Upper Mississippi Valley region.

Chatfieldian K-bentonite Beds. The Chatfieldian (Fig. 8) Dickeyville K-bentonite Bed (Willman and Kolata, 1978) has its type section in the upper part of the Guttenberg Limestone Member of the Decorah Formation 6.4 km northwest of Dickeyville, Wisconsin, and has been correlated by chemical fingerprinting methods through the Upper Mississippi Valley region (Kolata et al., 1986, 1987). The bed is present in a drill core near the top of the Guttenberg Limestone in Louisa County, Iowa (locality 96), 8 m above the Deicke K-bentonite. The bed is marked by a deflection on a gamma-ray log from nearby Washington County and can be traced on wireline logs southward into eastern Missouri (Plate 2). Based on the regional uniform thickness of the stratigraphic intervals between the Deicke, Millbrig, and Dickeyville K-bentonites, it appears that the Dickeyville is equivalent to the House Springs K-bentonite Bed in the basal part of the Kimmswick Limestone of eastern Missouri (Kolata et al., 1986). This correlation must be considered provisional, however, until more definitive subsurface information is available in the area between eastern Missouri and southwestern Wisconsin. In the Illinois Basin, the Dickeyville produces a prominent deflection on logs in Douglas, Piatt, and Champaign Counties, Illinois (Plate 3). In the Michigan Basin, some wireline logs show deflections in the lower part of the Trenton Limestone that may correspond to the Dickeyville, but this K-bentonite does not appear to be very thick or widespread in the basin.

The Dickeyville K-bentonite is traceable in the subsurface of the Illinois Basin from easternmost Missouri through Illinois, Indiana, and Kentucky into southern Ohio (Plates 4 and 5) where it correlates with a bed noted by Stith (1986) in the basal part of the Lexington Limestone in Butler County at 11 m above the Deicke K-bentonite. This interval thickens eastward to approximately 18 m in Highland County. South-southwestward in Campbell County, Kentucky (locality 191), the bed is present at 10 m above the Deicke, the interval apparently decreases southward in the area of the Jessamine Dome in north-central Kentucky. The Capitol "metabentonite," described by Conkin and Dasari (1986) from exposures of the Curdsville near Frankfort, Franklin County, Kentucky, very likely correlates with the Dickeyville.

Southeast of Highland County, Ohio, the Dickeyville is readily traced on wireline logs through the eastern counties of Kentucky into Lee County, Virginia (Plates 5 and 6). Within Lee County, the bed occurs in the Trenton Limestone at 32 m above the Deicke K-bentonite in the Shell Oil Co. No. 1 L. S. Bales well (locality 164) and 37 m above the Deicke in outcrop at Hagan, Virginia (locality 59). In Russell County, Virginia, the Gulf Oil Corp. No. 1 W. R. Price well (locality 163) penetrated the Pine Mountain thrust fault and encountered an overthrusted stratigraphic succession that is coextensive with the western outcrop belt of the Valley and Ridge Province. The Hardy Creek/Eggleston/Trenton succession is very similar to that in Lee County, Virginia. The interval between the Dickeyville and Deicke thickens to about 65 m. These correlations clearly demonstrate the greater thickness of the stratigraphic succession as a result of rapid subsidence during Middle Ordovician time in the Sevier foreland basin (Plate 6).

The Dickeyville is equivalent to K-bentonite bed R12 (Miller and Fuller, 1954; Miller and Brosgé, 1954) in the western outcrop belt of the Valley and Ridge Province. According to Haynes (1992, 1994), R12 is equivalent to K-bentonite V-7 (Rosenkrans, 1936) in the Bays Formation of the central outcrop belt.

Although the signature of a K-bentonite is locally present on wireline logs in Tennessee and Alabama in the stratigraphic interval containing the Dickeyville, the bed does not appear to be as thick or widespread as it is farther north. The T-5 K-bentonite (Wilson, 1949), occurring at 1 m to 3 m above the Carters Limestone in central Tennessee, is in approximately the same stratigraphic position as the Dickeyville, but there is not enough outcrop or subsurface data to be certain of this correlation.

The Elkport K-bentonite is known from outcrops in the Upper Mississippi Valley region where it occurs locally in the basal part of the Guttenberg Limestone Member (Willman and Kolata, 1978). It is present in some cores in eastern Iowa, but it does not produce a persistent and easily recognized deflection on wireline logs in the Mississippi Valley region because it commonly is <2 cm thick. The Elkport may be equivalent to one of two local K-bentonites in the Kings Lake Formation of eastern Missouri (Kolata et al., 1986). A persistent deflection on gamma-ray logs at the projected level of the Elkport in southern Ohio is traceable through eastern Kentucky into western Virginia (Plate 6). In outcrop at Hagan, Virginia (locality 59), the K-bentonite is 12 cm thick and occurs 10 m above the Millbrig K-bentonite in the Trenton Limestone. It produces a gamma-ray deflection at a level of 10 m above the Millbrig in the nearby Shell Oil Co. No. 1 L. S. Bales well (locality 164). The bed appears to correlate with K-bentonite R11 (Miller and Fuller, 1954; Miller and Brosgé, 1954) in the Rose Hill district of Lee County, Virginia. In the Ohio/Kentucky/Virginia region, we tentatively assign the name "Elkport" to this bed, but are awaiting additional outcrop and subsurface studies to confirm the correlation.

In the Mississippi Valley, 13 K-bentonite beds are known to occur above the Dickeyville (Fig. 11) in Mohawkian and Cincinnatian age rocks. Assuming that these beds represent the distal edges of volcanic ash layers derived from volcanoes situated along the eastern margin of North America, the beds should be traceable eastward into the Appalachian region. However, correlations are difficult to determine because the beds are relatively thin and somewhat discontinuous. Furthermore, in some regions these beds probably are not obvious on wireline logs because the shaly rocks that may contain them yield gamma-ray/neutron logs that show little contrast between the K-bentonites and confining strata. Also, no obvious thick and continuous K-bentonites have been found in the Trenton Group between northern Alabama and western Virginia. This suggests that middle to late Chatfieldian source volcanoes were situated north of the Carolinas, in contrast to the Turinian and early Chatfieldian beds such as the Hockett, Ocoonita, Deicke, Millbrig, and Dickeyville, that were derived from volcanoes situated between Alabama and North Carolina. The relatively large number of late Mohawkian (Chatfieldian) K-bentonites in West Virginia, northern Virginia, Pennsylvania, New York, and St. Lawrence lowlands of Ontario and Quebec suggests that most source volcanoes were situated in the region of New England and/or Maritime Canada.

Cincinnatian Series

The base of the Cincinnatian Series, as defined in the standard reference section from a continuous 368 m core drilled by Cominco American, Inc. just east of Minerva, Mason County, Kentucky, is situated at a point in the upper part of the Point Pleasant Formation (Fig. 24) as determined by conodont-based graphic correlation methods (Sweet, 1995). Graphic correlation methods have been used to identify this stratigraphic level in other parts of eastern North America. In the upper Mississippi Valley region the base of the Cincinnatian is situated at approximately the contact between the Dunleith and Wise Lake Formations (Fig. 11) close to the level of the Dygerts K-bentonite (Sweet, 1987). The Dygerts K-bentonite, having its type section in the Wise Lake Formation in northern Illinois (Willman and Kolata, 1978), appears to have good potential for regional correlation, and may have some utility in identifying the series boundary. As noted earlier in this book, the Dygerts is widespread in outcrop and the subsurface of the Upper Mississippi Valley. Traced eastward into the Michigan Basin, the Dygerts appears to correspond to the persistent deflection on gamma-ray logs from the upper part of the Trenton Limestone (Plate 3). Although not identified as the Dygerts, Travail (1990) observed this bed in cores and wireline logs in the Sherman Falls Formation of southwestern Ontario. A possible equivalent K-bentonite is present in the same stratigraphic interval in the upper part of the Neuville Formation in the St. Lawrence lowlands (Brun and Chagnon, 1979). The southern extent of the Dygerts is unclear, particularly in the Illinois Basin where the Galena is relatively thin. Additional stratigraphic studies are needed to determine if the Dygerts is equivalent to the K-bentonite exposed 20 m below the top of the Galena Group near Clarksville, Missouri (locality 41), and the K-bentonite interpreted from wireline logs in the upper part of the Galena in Pike, Douglas, Piatt, Williamson, Wayne, and Lawrence Counties, Illinois (Plates 2, 3, and 4).

The Westboro and Bear Creek K-bentonite zones (Schumacher and Carlton, 1991) in the Lexington Limestone and Point Pleasant Formation are approximately the same age as the K-bentonites in the middle and upper Galena Group of the Upper Mississippi Valley. Likewise, the cluster of K-bentonites in the Dolly Ridge Formation of West Virginia (Perry, 1972; Ryder, 1992b), upper Trenton Group and Utica Shale in New York State (Goldman et al., 1994; Mitchell et al., 1994), and Montreal and Neuville Formations of the St. Lawrence lowlands (Brun and Chagnon, 1979) are probably in part equivalent to those in the middle and upper Galena. Precise correlations are yet to be established.

A K-bentonite in the Maquoketa Group (Cincinnatian Series undifferentiated) in the Seneca County, Ohio, drill core (Ohio Division of Geological Survey core 2580) may correspond to one of the K-bentonites in the Dubuque Formation (Bergström and Mitchell, 1992) or possibly to the bed in the Maquoketa Group in Fillmore County, Minnesota. The post–*G. pygmaeus* age of the Ohio bed is suggestive of a correlation with the upper *A. superbus* Zone–lower *A. ordovicicus* Zone beds in Iowa and Minnesota. The two beds reported by Willman and Templeton (1963) from the Scales Formation, Maquoketa Group, at Kentland, Indiana, may also correspond to this group of K-bentonites.

REGIONAL CROSS SECTIONS

The Middle Ordovician succession of the eastern Midcontinent is shown on six regional cross sections based primarily on wireline logs (Plates 2, 3, 4, 5, 6, 7). The K-bentonites provide insight to the temporal relations of these strata. Additional observations and comments on each cross section are summarized below.

Eastern Missouri–southeastern Minnesota cross section

Plate 2 shows the stratigraphic relations in a cross section extending from St. Charles County, Missouri, northward along the western margin of the Illinois Basin through the subsurface of Missouri, Illinois, and Iowa to Freeborn County, Minnesota.

1. The Deicke, Millbrig, and Dickeyville K-bentonites are widely present in eastern Missouri, western Illinois, eastern Iowa, and southern Minnesota in an interval, no more than 10 m thick, consisting of the Decorah Formation and basal part of the

Galena Group. The interval is amazingly uniform in thickness and lithologic character along this transect, a distance of about 800 km. The time relations established by the K-bentonite beds indicate that some stratigraphic units are in part laterally gradational yet remarkably uniform in terms of depositional rates and sedimentary environments.

2. In eastern Missouri, the Deicke occurs at the base of the relatively pure Castlewood Limestone Member of the Decorah Formation (Fig. 18). Although lithologically similar to the underlying Plattin Group, Templeton and Willman (1963) chose to assign the Castlewood to the Decorah because it is more argillaceous, coarser grained, more massive and lacks the conchoidal fractures of the underlying limestone. The Castlewood, readily recognized on gamma-ray logs, is uniform in thickness (approximately 1 m) and lithology from Missouri to southern Minnesota where it is referred to as the Carimona Limestone (Weiss, 1955).

3. The Millbrig is present in the Spechts Ferry Shale Member of the Decorah Formation. The shale is approximately 1.5 m thick near St. Louis, Missouri, diminishes to 0.3 m of argillaceous dolomite in Hancock County, Illinois, and thickens to 15 m in Freeborn County, Minnesota, where the Guttenberg Limestone and basal carbonate rocks of the Galena Group grade to shale. Thickness patterns suggest that the Spechts Ferry Shale was derived from the uplifted and exposed igneous terranes on the Transcontinental Arch and probably from reworked Cambrian shales and local exposures of granite and rhyolite on the uplifted Ozark Dome.

4. The Dickeyville K-bentonite, formerly referred to as the House Springs K-bentonite (Kolata et al., 1986), occurs 1 m above the base of the Kimmswick Limestone in eastern Missouri. Traced northward on wireline logs and in drill cores it appears to cross into the upper part of the Guttenberg Limestone north of Louisa County, Iowa. Assuming this correlation is correct, then the basal Kimmswick grainstone facies of eastern Missouri is coeval with the Guttenberg lime-mudstone facies of the Upper Mississippi Valley. These stratigraphic relations also show that the Kings Lake Formation, recognized by Templeton and Willman (1963) in eastern Missouri, is a silty carbonate facies of the Spechts Ferry Shale and the Guttenberg Limestone.

5. The Elkport K-bentonite appears to be less widespread than the Deicke, Millbrig, or Dickeyville. It is present locally in the subsurface of Clayton County, Iowa, but does not produce a noticeable deflection on wireline logs. A 2 cm to 3 cm thick bed present locally in the Kings Lake Formation of eastern Missouri may be equivalent to the Elkport (Kolata et al., 1986).

6. The Plattin Limestone of Missouri and its equivalent, the Platteville Group of the Upper Mississippi Valley, thickens southward from approximately 2 m in Freeborn County, Minnesota, to 50 m in St. Charles County, Missouri. The Platteville is particularly thin in Hancock County, Illinois, and Clark County, Missouri, an area that lies along the crest of the Sangamon Arch, suggesting that the arch was uplifted during deposition of the Platteville, prior to deposition of the Deicke K-bentonite. The structure marks the boundary between the shoreward lagoonal Glenwood Shale to the north and the marine carbonates of the Joachim Formation to the south (Treworgy et al., 1994). Further, the arch underwent major uplift during the Late Silurian and Early Devonian, resulting in truncation of the Galena Group and deposition of the Middle Devonian Wapsipinicon Limestone.

Eastern Missouri–southern Ontario cross section

Plate 3 illustrates the stratigraphic relations in a cross section extending from Perry County, Missouri, northeastward through the Illinois Basin across the Kankakee Arch and through central Michigan to Bruce County, Ontario, on the northeastern side of the Michigan Basin.

1. Wireline logs and drilling samples show that the Spechts Ferry Shale and Kings Lake Limestone Members of the Decorah Formation diminish in thickness between Perry County, Missouri, and Perry County, Illinois, a distance of about 50 km. Consequently, the Decorah Formation is not a practical nor useful unit to trace in the Illinois Basin. In addition, although the Castlewood Limestone Member (Fig. 18) is easily recognized as the limestone unit between the Deicke and the overlying Galena Group, it has not been utilized in most subsurface studies in the basin. The contact between the Platteville (Black River) and overlying Galena (Trenton) is generally identified as the upward change from relatively pure, micritic limestone to relatively argillaceous grainstone and wackestone. On gamma-ray logs, the contact is marked by a slight to moderate increase in natural radiation in the basal argillaceous strata of the Galena. Typically, the Millbrig K-bentonite occurs at or immediately below the Galena-Platteville contact. Following common practice of geologists working the subsurface of the Illinois Basin, the Decorah is terminated by vertical cutoff along the western margin of the basin, and the Castlewood is included in the Platteville.

2. The Ocoonita, Deicke, Millbrig, and Dickeyville K-bentonites are particularly well developed in Douglas, De Witt, Piatt, and Champaign Counties in east-central Illinois, suggesting that local depositional environments were conducive to the preservation of volcanic ash. This is not the case a short distance to the north in Iroquois County, Illinois, where only the Deicke can be identified confidently on wireline logs.

3. The Middle Ordovician stratigraphic succession changes significantly from Iroquois County northward through Lake County, Indiana, to Berrien County, Michigan, reflecting uplift of the Kankakee Arch (Atherton, 1971). Truncation of the Early Ordovician Prairie du Chien Group along the crest of the arch indicates an episode of uplift and erosion occurring between late Ibexian and late Whiterockian time, concurrent with development of the regional sub-Tippecanoe unconformity (Kolata, 1991). Truncation of the St. Peter Sandstone and overlying Joachim Formation suggests a second period of uplift during early Turinian time. A third episode of uplift occurring in late Turinian time is apparent from thinning of the Platteville (Black River) Group on the flanks and crest of the arch. Furthermore,

regional stratigraphic relations suggest that the uppermost units of the Platteville, including the Deicke and Millbrig, were eroded from the crest of arch prior to deposition of the Galena Group. This is the case in the U.S. Steel Co. No. 1 Gary Sheet and Tin Plant well in Lake County, Indiana (locality 131). The axis of the arch migrated at least 50 km southwestward during Whiterockian and Turinian time as indicated by the location of maximum truncation during the three episodes of uplift. Regional stratigraphic relations also indicate that the Kankakee Arch was structurally continuous with the Wisconsin and Cincinnati Arches during some episodes of Ordovician deformation.

4. The Trenton Limestone is approximately five times thicker in the Michigan Basin than in the Illinois Basin. Measured intervals between K-bentonite beds and trends in the vertical distribution of argillaceous material, as indicated on gamma-ray logs, suggest that sedimentation rates were considerably higher in the Michigan Basin during deposition of the Trenton. It is possible that the upper part of the Galena in the Illinois Basin grades to shale in the lower Maquoketa Group, but the Maquoketa does not show a reciprocal increase in thickness.

Eastern Missouri–eastern Ohio cross section

Plate 4 shows the stratigraphic relations in a cross section extending from the eastern flank of the Ozark Dome in Perry County, Missouri, across the Illinois Basin, Sebree Trough, and Cincinnati Arch to Noble County in eastern Ohio.

1. The thickness and lithologic character of the Black River (=Plattin=Platteville) strata are amazingly uniform along the 600 km transect. This is particularly evident in the K-bentonite–bearing interval as seen on the wireline logs. The upper Black River consists of micritic limestone extending from the outcrop belt in eastern Missouri to the subsurface of eastern Ohio. Furthermore, regional studies show the Black River to be one of the most homogeneous and widespread Paleozoic units in North America. The lower part of the Black River is somewhat thicker in the southern part of the Illinois Basin. Wireline logs show the interval between the Hockett and Deicke K-bentonites to be relatively thick in Hamilton County, Illinois, but thinning gradually eastward toward the Cincinnati Arch.

2. The Trenton is unusual because it is relatively thin (~30 m) in the region of Hamilton County, Illinois, an area where the underlying Black River is relatively thick. Assuming that the apparent K-bentonite bed in the upper part of the Trenton on the logs from Williamson and Wayne Counties, Illinois, is equivalent to the interpreted K-bentonite on logs from Lawrence County, Illinois, and Vigo, Clay, and Green Counties, Indiana, it would appear that sedimentation rates decreased dramatically in the Hamilton County region during late Mohawkian time. A part of the upper Trenton may grade to shale in the overlying Maquoketa Group in the Hamilton County area, but this probably is not significant because the Maquoketa is not reciprocally thickened. In fact, isopachs of the Maquoketa show it to be uniform in thickness at approximately 60 m from Hamilton County to northeastern Iowa (Willman and Buschbach, 1975). It also seems unlikely that the basin was uplifted and eroded after deposition of the Galena as suggested by Templeton and Willman (1963). On the contrary, many researchers believe that there was a rise in sea level accompanied by a transition from relatively oxygen-rich (Galena) to oxygen-poor (Maquoketa) conditions (Keith, 1985; Witzke and Kolata, 1988; Bergström and Mitchell, 1992). Sea-level rise and deepening very likely would have affected deposition in the basins before rising over the surrounding arches and domes. Conodonts processed from the Superior Oil Co. No. C-17 Ford et al. drill core (locality 119) in nearby White County, Illinois, show the top of the Galena to be no younger than *P. tenuis* (Thomas H. Shaw and Walter C. Sweet, 1994, written communications), indicating a late Mohawkian age. In contrast, north of the Illinois Basin in the Upper Mississippi Valley region, the uppermost part of the Galena contains *A. ordovicicus* Zone conodonts of late Cincinnatian age, as does the Cape Limestone situated at the top of the Galena in eastern Missouri (Mackenzie and Bergström, 1993). This indicates that the top of the Galena is diachronous, becoming younger away from the basin center. These stratigraphic relations suggest that a starved basin existed in southern Illinois during deposition of the Trenton and Maquoketa. The late Mohawkian through Cincinnatian depositional history of the basin, however, is poorly known because of the paucity of faunal data from the Trenton and Maquoketa.

3. Between Decatur County, Indiana, and Butler County, Ohio (localities 180 and 181), the Galena (Trenton) grades to shale of the Maquoketa Group (Utica Shale) in the enigmatic Sebree Trough (Keith, 1985, 1989). This narrow shale trough is believed to have developed during Chatfieldian time in western Tennessee and Kentucky. By late Chatfieldian time the trough connected to the subsiding Martinsburg Basin (Fig. 1) in eastern Pennsylvania and apparently to open ocean south and west of Tennessee (Keith, 1989). The trough separates the relatively pure limestone and dolomite of the Galena to the northwest from the argillaceous Lexington Limestone to the southeast. The facies relations are established by graptolite zonation (Bergström and Mitchell, 1992). Stratigraphic relations suggest that the Sebree Trough developed shortly after deposition of the Millbrig K-bentonite.

4. Wireline logs in northern Kentucky, southeastern Indiana, and southwestern Ohio indicate that the interval of Tyrone and Lexington Limestone between the Millbrig and Dickeyville K-bentonites gradually increases from 6 m to 12 m between Switzerland County, Indiana, and Clinton County, Ohio. Stith (1986) suggested the same correlation and stratigraphic relations based on subsurface data from southwestern Ohio and northern Kentucky.

5. East of Fairfield County, Ohio, the K-bentonites are somewhat thinner, intermittent in occurrence, and a challenge to correlate.

Eastern Missouri–western Virginia cross section

Plate 5 shows the stratigraphic relations in a cross section extending from the eastern flank of the Ozark Dome in St.

Charles County, Missouri, across the Illinois Basin, Sebree Trough, and Cincinnati Arch to the margin of the Sevier Basin (Fig. 1) in Russell County, Virginia.

1. The western part of this section shows the southward increase in thickness of the Decorah Formation and associated K-bentonite–bearing interval in southeastern Missouri and southwestern Illinois toward the Illinois Basin depocenter situated in southernmost Illinois and western Kentucky (Kolata, 1991). Thickening of the Castlewood Limestone is particularly notable. As noted below, the Castlewood is included in the upper part of the Platteville in the Illinois Basin and correlated with the upper Tyrone Limestone in Kentucky, upper Carters Limestone in Tennessee, and the upper Eggleston Limestone in Virginia.

2. The eastern part of this cross section illustrates the correlation of K-bentonites from the Sevier Basin (Fig. 1) westward across the Cincinnati Arch showing the equivalent beds, b, a, γ, β, and α, described by Stith (1979, 1986). Accordingly, the following correlations are demonstrated: (1) the Hockett is equivalent to b and can be traced from the Sevier Basin into the Illinois Basin; (2) bed a is thin and not readily traceable; (3) the Ocoonita is equivalent to γ and traceable into the Illinois and Michigan Basins; (4) the Deicke is equivalent to β (Pencil Cave); (5) the Millbrig is equivalent to α (Mud Cave); and (6) the Dickeyville was recognized by Stith (1986) but not distinguished with a symbol or name. The intervals between the K-bentonites thicken gradually from the Cincinnati Arch to the margin of the Sevier Basin in Virginia where they are approximately twice as thick.

3. Throughout the Upper Mississippi Valley region (Plate 2), the Millbrig lies within the Decorah Formation, Galena Group (=Trenton), but in the area of southwestern Ohio, as throughout most of the Illinois Basin, the Millbrig lies in the upper part of the Tyrone Limestone (=Black River Group), overlain by as much as 2 m of pure micritic limestone (Stith, 1986). This is significant because in the Mississippi Valley the top of the Carimona Limestone/Castlewood Limestone/Platteville Group (=Black River Group) is a sharp break between micritic limestone below and shale or pure to argillaceous grainstone of the Trenton Group above. In parts of southern Ohio, however, the prominent boundary between the Tyrone and Lexington Limestones lies above the Millbrig K-bentonite. For example, in the ODS-2621 drill core from Highland County, Ohio (locality 189), a hardground at the top of the Tyrone Limestone occurs at a depth of 986 ft (300.5 m), 1.5 m above the Millbrig K-bentonite (Gregory A. Schumacher, 1994, written communication). The basal 12 cm of the overlying Lexington Limestone contains lithoclasts of the Tyrone Limestone. Likewise, in drill core ODS-3019 in Hamilton County, Ohio, the top of the Tyrone, 1 m above the Millbrig, is a sharp, undulatory contact with pyrite-filled borings that penetrate 3 cm into the Tyrone, typical of a hardground. These stratigraphic relations indicate that the micritic limestones in the upper part of the Tyrone were deposited in southern Ohio at the same time that basal Trenton shales were deposited in the Mississippi Valley. Furthermore, the hardground is either diachronous, being older in the central Midcontinent United States than it is east of the Cincinnati Arch or, more likely, there are multiple hardgrounds. Diachronous hardgrounds, resulting from sea-level rise, are forming presently in the Persian Gulf (Purser, 1969). From the perspective of sequence stratigraphy, this is not insignificant because sequence boundaries are commonly placed at prominent lithologic breaks interpreted to be time planes that define the chronostratigraphic framework. This is particularly relevant to the concept of paracontinuous stratigraphy as applied to the Ordovician succession of eastern North America (Conkin and Conkin, 1973; Conkin, 1991). These authors interpret the Turinian/Chatfieldian stage boundary to be a discontinuity of regional chronostratigraphic significance, extending "over an area of more than one million [square] miles in eastern United States and southern Ontario" (Conkin, 1991, p. 1). Their approach assumes that physical discontinuities, such as hardgrounds, parallel time lines and can be correlated regionally. This line of reasoning has been used by Conkin (1991) to challenge our correlation of K-bentonite beds. Using outcrop and subsurface data, we argue that K-bentonites provide time resolution that is much more precise and less ambiguous than that determined from tracing of hardgrounds and small-scale unconformities between widely spaced outcrops. Our stratigraphic analysis includes a study of outcrops and several hundred wireline logs linking the subsurface with surface exposures.

Southern Ohio–western Virginia cross section

Plate 6 shows the stratigraphic relations in a cross section extending from the Cincinnati Arch in Highland County, Ohio, to the margins of the Sevier Basin in Russell County, Virginia. The transect forms a stratigraphic tie between cross sections (Plates 4, 5, and 7).

1. The Highland County, Ohio, well shows the primary log characteristics, including K-bentonites and shaly zones, used by Stith (1979, 1986) to correlate the upper Black River and lower Trenton strata of southwestern Ohio. The petrophysical response is significant for K-bentonites b (Hockett), γ (Ocoonita), β (Deicke), and α (Millbrig). K-bentonite bed a is not widespread and persistent. Marker bed Δ is a persistent, non–K-bentonitic, argillaceous unit that is widespread in the region of the Cincinnati Arch and Cumberland Plateau.

2. The Hockett K-bentonite occurs in the upper part of the Hardy Creek Limestone in Virginia and in the upper Tyrone Limestone in Kentucky and Ohio. The Ocoonita, Deicke, and Millbrig are present in the Eggleston Limestone and the Dickeyville and possibly the "Elkport" in the lower part of the Trenton Limestone.

3. The stratigraphic section in easternmost Kentucky and western Virginia is notable because of the increased thickness of strata in the K-bentonite–bearing interval; this indicates relatively rapid subsidence in the Sevier foreland basin during Turinian and Chatfieldian time coincident with the late stage of

the Blountian tectophase (Ettensohn, 1991) of the Taconian orogeny.

Northeastern Kentucky–northeastern Alabama cross section

Plate 7 shows the stratigraphic relations in a cross section extending from Greenup County, Kentucky, southwestward across the Cumberland Plateau to Jackson County, Alabama.

1. This cross section illustrates the correlation of the Deicke and Millbrig K-bentonites with beds T-3 and T-4 (Wilson, 1949) and the probable correlation of the Ocoonita with T-2 and the Dickeyville with T-5.

2. In northeastern Kentucky and adjacent parts of southeastern Ohio, the K-bentonites are relatively thin and intermittent in occurrence. An analysis of wireline logs in this region suggests that the Millbrig may not be present, perhaps removed by pre-Lexington erosion. South of Montgomery County, Kentucky, however, the very prominent deflections of the Deicke, Millbrig, and Dickeyville gamma-ray/neutron logs suggest that the beds are relatively thick and that the transect crosses the long axis of the original ash deposits.

3. Between drill holes in Montgomery and Lincoln Counties, Kentucky, there is a gradual increase in thickness of the interval between the Deicke and the top of the Tyrone Limestone. This interval remains relatively thick through northern Tennessee, but decreases slightly in Franklin County, Tennessee, and Jackson County, Alabama.

TECTONIC SETTING AND ORIGIN OF K-BENTONITES

It is believed that the Taconian Orogeny resulted from continental margin–island arc collisions associated with east-dipping subduction zones during the Middle and Late Ordovician (e.g., Malpas and Stevens, 1977; Hiscott, 1978; Rowley and Kidd, 1981; Rodgers, 1987; Hatcher, 1989). There is a growing trend, however, to differentiate between the various episodes of Taconian deformation that occurred along the eastern margin of Laurentia (Drake et al., 1989). The temporal and spatial distribution of Ordovician deep foreland basin sequences in the central and southern Appalachians and along the western margin of the northern and Maritime Appalachians suggests that the orogeny occurred in several diachronous tectophases related to different convergence centers (Shanmugam and Lash, 1982; Diecchio, 1991; Ettensohn, 1991). The foreland basins are believed to have formed by thrust-loading during subduction (Jacobi, 1981; Quinlan and Beaumont, 1984; Lash, 1987). Furthermore, evidence suggests that these tectophases were characterized by volcanic activity resulting in plinian eruptions that were the source of the Ordovician K-bentonite beds found in eastern North America. Locations of source volcanoes and sequence of volcanic events can be inferred from age, chemical composition, thickness, grain size, and distribution of K-bentonites.

One of the earliest phases of Taconian subduction and associated foreland basin development appears to have occurred during early Middle Ordovician time in Newfoundland (Diecchio, 1991). The stratigraphic sequence represented by the Table Head Group is interpreted to have been deposited in a deep foreland basin apparently associated with thrust-loading during Whiterockian time (Jacobi, 1981; Cawood et al., 1988). Occurrence of the stratigraphically and geographically isolated K-bentonites in the mid-Whiterockian (*P. tentaculatus* Zone) Table Head and Cape Cormorant Formations (Fig. 64) in western Newfoundland (Plate 1) suggests that the source volcanoes were associated with a nearby subduction system.

In the southern Appalachians, collision and subsequent docking of a deformational load against the Laurentian platform margin initiated the Blountian tectophase (Ettensohn, 1991). The colliding terrane is referred to as the Carolina terrane (Secor et al., 1983; Hatcher, 1989) or Oliverian-Midland Valley island arc (McKerrow et al., 1991). Although the initial collision is believed to have begun as early as the Cambrian (Hatcher, 1987), development of the deep Sevier foreland basin and deposition of the associated black shales did not take place until late Whiterockian time. The K-bentonite beds in the Athens and Blockhouse Shales, as well as in the Little Oak Limestone (Fig. 33), were likely derived from a volcanic arc associated with the Blountian tectophase. Likewise, the early and mid-Mohawkian K-bentonites, including the Hockett through Dickeyville beds (Hagan K-bentonite Complex), are thicker and coarser toward the southern Appalachians (Plate 6), suggesting that source volcanoes were on an island arc situated between Alabama and South Carolina. The most voluminous ash beds, including the Deicke and Millbrig, were deposited during the late stages of collision and basin development. Chemical analyses indicate that the K-bentonites were derived from calc-alkaline, destructive plate-margin volcanics. Furthermore, phenocryst grain-size data from the Millbrig suggest that the source volcanoes were situated in the southern Appalachian region (Zhang and Huff, 1994).

As the Sevier basin reached its final cycle of development, the Martinsburg Foreland Basin (Fig. 1) in the central Appalachians began to form in response to a new episode of deformational loading centered near the New York promontory (Shanmugam and Lash, 1982; Ettensohn, 1991). This marked the inception of the late Mohawkian Taconic tectophase. The northward migration of tectonic activity was accompanied by a corresponding shift in the location of volcanic centers, as suggested by the distribution of late Mohawkian and early Cincinnatian K-bentonite beds. The numerous ash beds in the Utica Shale of New York and southern Quebec appear to have been derived from an island arc east of the central and/or northern Appalachians. The apparent proximity of the Quebec Appalachians of Canada to nearby volcanism is indicated by the presence of 10 m to 30 m thick beds of tuff and lapilli tuff in the predominantly graptolite-bearing (*D. multidens* Zone), nonvolcanic turbidite sequences of the Middle Ordovician Saint-Victor

Formation (Cousineau, 1994). These arc-derived volcaniclastic rocks are believed to have been deposited in a deep and narrow fore-arc basin during the Taconian tectophase.

Supporting evidence for interpretations of the tectonic setting and origin of the Ordovician K-bentonites of eastern North America comes from the use of magmatic and tectonic discrimination diagrams. These are empirically derived, but nonetheless quite instructive methods of examination of geochemical data which are widely used in igneous petrology (Wang and Glover, 1992).

Composition of parental magmas

We have analyzed approximately 750 whole-rock samples of Ordovician K-bentonites for both major and trace elements in order to gain a better understanding of their regional compositional variation. Certain trace elements including the large ion lithophile (LIL) and rare earth elements (REE) are considered to be immobile during the course of devitrification of volcanic ash and thus offer the best sources of information concerning parental magma composition. In addition, we have acquired some preliminary information on the major element composition of pristine melt inclusions in primary crystals. The analyzed samples range from Whiterockian to Cincinnatian in age, although most are Mohawkian, and the data generated by these studies have formed the basis for interpretations of tectono-magmatic setting.

Data for 138 samples were plotted on the Zr/TiO_2 versus Ga magmatic discrimination diagram of Winchester and Floyd (1977; Fig. 69 herein). The K-bentonite samples belong to a calc-alkaline suite ranging from andesites through rhyodacites and trachyandesites to rhyolites. Samples generally have <30 ppm Ga, which is characteristic of subalkaline magmas, whereas alkaline magmas tend to show an increased concentration of Ga with differentiation (Winchester and Floyd, 1977). Similarly, the Zr/TiO_2 ratio increases with progressive differentiation of a basaltic magma, reflecting the overall decline of TiO_2 in intermediate and felsic rocks. No noticeable evolutionary trends can be detected in the data when examined in stratigraphic succession, and thus there is no evidence from these data of a continuing or systematic development of collision margin volcanism through the time interval represented by the K-bentonite beds.

Rare earth element patterns for two of the most prominent K-bentonite beds in North America, the Deicke and Millbrig, show broadly similar features. Chondrite-normalized REE plots of Deicke and Millbrig samples (Fig. 70) show typical subalkaline and peralkaline patterns (Roberts and Merriman, 1990), and both show some evidence of light rare earth element (LREE) enrichment characteristic of highly evolved calc-alkaline magmas. The Millbrig patterns show somewhat less LREE enrichment and may represent a slightly less fractionated magma. However, both beds have a pronounced negative Eu anomaly indicating derivation from a highly evolved magma in which significant plagioclase fractionation occurred prior to the episode of volcanism. It would seem unlikely that the source magmas were derived by direct fractionation from an oceanic or upper mantle source, but more likely that they were produced by partial melting of continental crustal rocks and possibly oceanic sediments during collision, as indicated by the LREE enrichment.

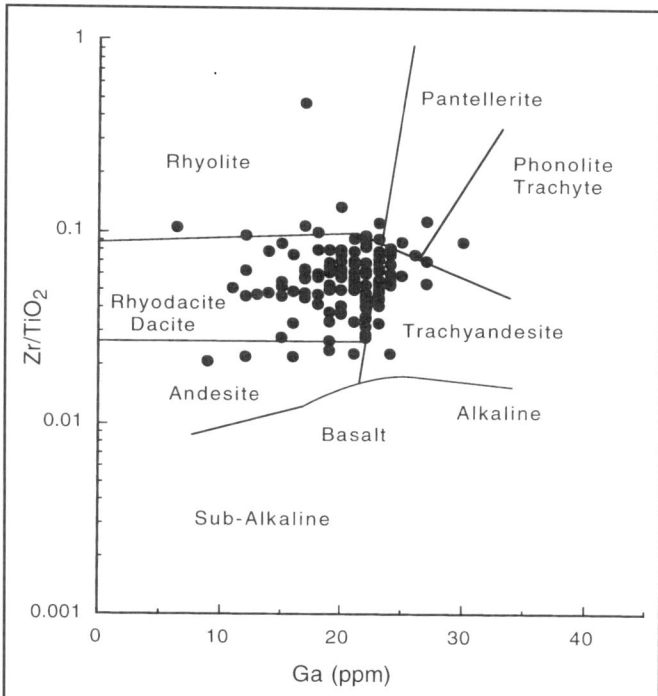

Figure 69. Data from Ordovician K-bentonite analyses plotted on the magmatic discrimination diagram of Winchester and Floyd (1977). Ratios of the immobile trace elements Ga, Zr, and Ti indicate most beds were derived from dacitic parent magmas.

La enrichment compared with Nb/U ratios supports the involvement of a dominantly crustal source. Figure 71 compares mid-oceanic ridge basalts (MORB) and continental crust values for Th/La versus Nb/U, and K-bentonites can be seen to correlate strongly with continental values. Further evidence for this comes from a comparison of Pb concentrations and Pb/Ce ratios for 190 K-bentonite samples (Fig. 72). When compared with oceanic basalts and oceanic island basalts versus deep-sea sediments, the Pb and Ce values show a ratio similar to sediments and, by analogy, the continental crust. Most studies of arc systems indicate that the involvement of subducted sedimentary components is limited to a few percent, and Figure 72 taken from the work of White (1989) also shows that additions of less than a few percent sediment to mantle or oceanic basalts leads to Pb/Ce ratios that are higher than any reported for fresh oceanic basalts. The close correlation of K-bentonite Pb-Ce values with oceanic sediments leads to the conclusion that melting of continental crustal rocks played a dominant role in the generation of the parent magmas.

Preliminary studies of melt inclusions in apatite phenocrysts provide additional information on the bulk composition

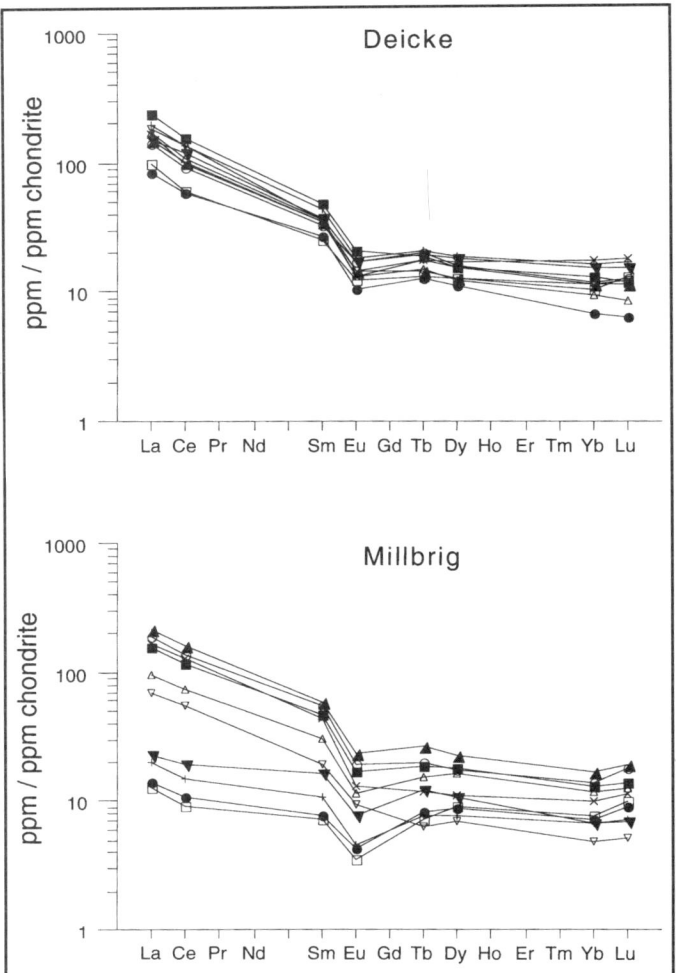

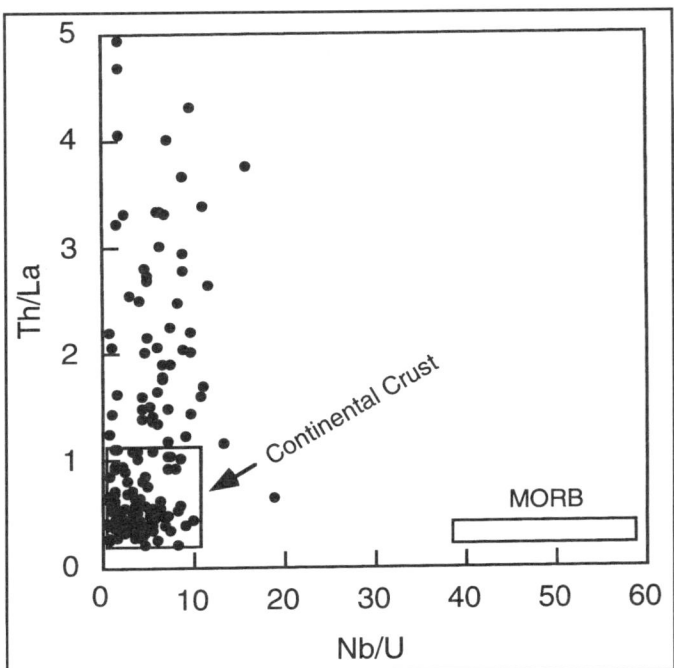

Figure 71. Bivariate plot of Ordovician K-bentonite samples using Nb-U versus Th-La. In contrast to mid-oceanic ridge basalt (MORB) samples, which have high Nb and low Th values, K-bentonite samples are similar to continental crustal rocks. Some K-bentonites have high La enrichment suggesting possible additional recycling of continental-derived sediments.

Figure 70. Chondrite-normalized rare earth element (REE) plots for 10 Deicke and 10 Millbrig samples from the southern Appalachians. The enrichment of light rare earth elements (LREE) and prominent negative Eu anomaly typify highly evolved calc-alkaline magmas and suggest a major component of continental crustal rock.

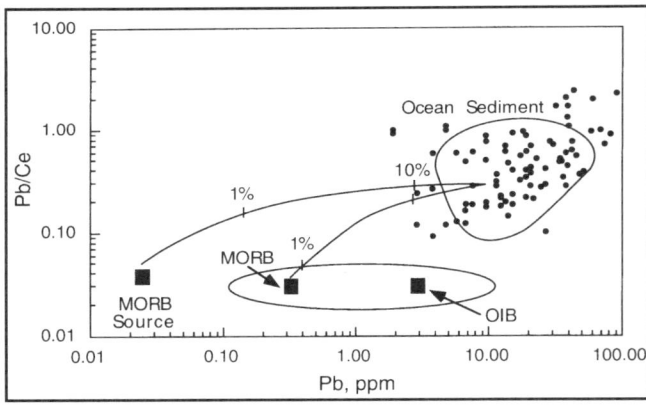

Figure 72. Pb versus Pb-Ce plot after White (1989) comparing oceanic sediment (continental crust) characteristics with oceanic basalts. The percent marks indicate that as little as 1% mixing of oceanic sediment with oceanic basalts would lead to Pb-Ce ratios in the mixture that are higher than any previously observed. These data suggest not only that very little mixing of sediment occurred in the upper mantle, but that K-bentonites have Pb and Ce values that can only be explained by a dominantly continental crustal source. Abbreviations: mid-ocean ridge basalts, MORB; and oceanic island basalts, OIB.

of the parent magma at the time of eruption. Major-element probe data from five inclusions from the Millbrig and Deicke K-bentonites, and the Kinnekulle K-bentonite for comparison, were analyzed by electron microprobe and the results plotted on an AFM diagram (Fig. 73). The glass is rhyolitic in composition with an estimated H_2O content between 4% and 5%. This value is derived by calculating normative compositions and using the experimental data of Lipman (1966), which relate the minimum melting composition to the solubility of water at a given pressure. The calculated pressures lie between 1,100 and 1,200 bars, which is equivalent to a depth of 4 km to 5 km to the magma chamber. It is interesting to note that total sulfur, expressed as SO_2, averages 0.038 wt%, which is about twice the petrographic value reported for the 1991 eruption of Mt. Pinatubo (Westrich and Gerlach, 1992). We estimate the minimum volume of sulfur injected into the stratosphere during the

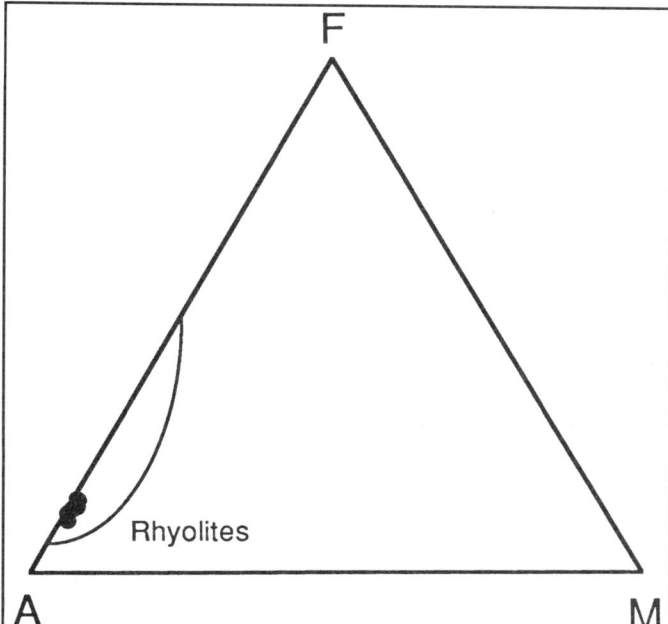

Figure 73. Microprobe analyses of pristine glass inclusions in apatite crystals from Ordovician K-bentonites plotted on an AFM diagram (A = alkalies, $Na_2O + K_2O$; M = MgO; F = FeO) fall well within the rhyolite field of Hildreth et al. (1991) for Yellowstone rhyolites. These data are more precise indicators of parental magma composition than whole-rock K-bentonite analyses.

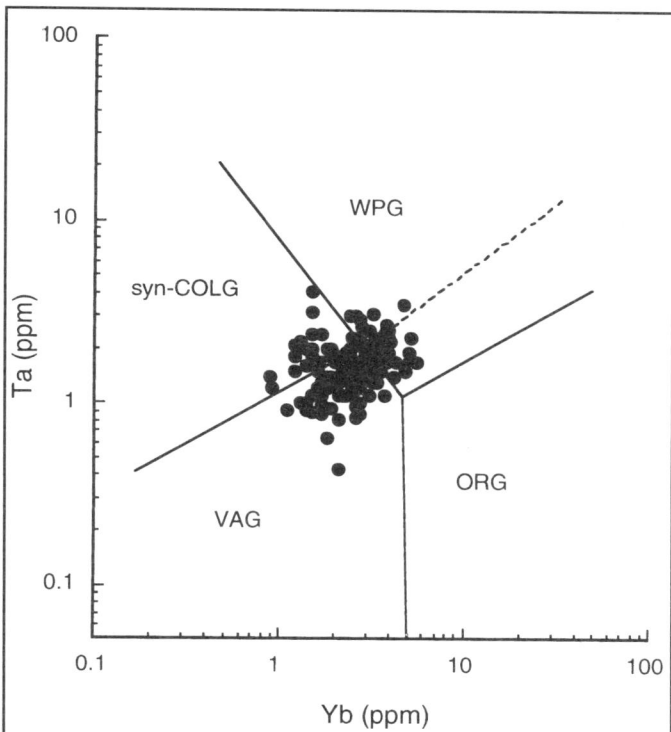

Figure 74. Data for 142 Ordovician K-bentonite samples plotted on the granite (G) tectonic discrimination diagram of Pearce et al. (1984) range from arc and syn-collision (syn-COL) to within-plate (WP) settings. Stratigraphic ranges of the samples do not indicate systematic change from one setting to another with time, and it is concluded that the plate setting of the source volcanoes includes both a subduction and plate margin component. Abbreviations: oceanic ridge granite, ORG; volcanic-arc granite, VAG.

Millbrig eruption was 59 Mt of elemental S equivalent to 118 Mt of SO_2. Much more may actually have been produced in the form of exsolved vapor phase SO_2 (Westrich and Gerlach, 1992). The conversion to sulfate aerosol would have yielded approximately 177 Mt of H_2SO_4 droplets. This is one to three orders of magnitude larger than any previously reported historic or pre-historic eruption and may have been capable of inducing rapid climatic change through alteration of the Earth's radiation budget (Rampino and Self, 1993).

Tectonic setting

The elements Nb, Ta, Yb, and Y are generally considered to be among the most alteration-independent of the immobile trace elements and were found by Pearce et al. (1984) to be particularly effective in discriminating between granites of diverse tectonic settings. Compared with volcanic-arc granites, Yb is more abundant in ocean ridge and within-plate granites, whereas Ta is particularly enriched in syn-collision and within-plate granites. Data for 142 North American Ordovician K-bentonite samples are plotted on the Ta-Yb discriminant diagram (Fig. 74). Most samples form a fairly tight cluster that falls on the junction between the volcanic-arc field, the syn-collision field, and within-plate field. The overall distribution pattern suggests some transition from arc or syn-collision to predominantly within-plate source magmas. Within-plate magmas are assumed to be derived from upper mantle sources where enrichment of Ta is related to the genesis of oceanic island–type basaltic magmas (Cheng and Pettry, 1993). The degree of depletion of the mantle source exerts control on the boundary between within-plate and oceanic ridge granites as does the early crystallization of crystalline phases such as magnetite and amphibole. Pearce (1982) used Zr/Ti ratios to discriminate between arc lavas and within-plate lavas. Further, felsic calc-alkaline magmas are typically enriched in LIL elements including Zr and thus can be distinguished from basalts formed in either environment. Ordovician K-bentonites plot dominantly in the arc lava field but extend into the within-plate field as well (Fig. 75). This distribution pattern reflects not only the felsic nature of the parental magmas but the strong involvement of continental crustal rocks. It would seem the volcanoes that produced the K-bentonite ash were characterized by both plate and island arc magmatic features and may well have been large calderas on the active margin of Laurentia. If the K-bentonites were arc derived, then there was a substantial amount of continental crust forming the basement.

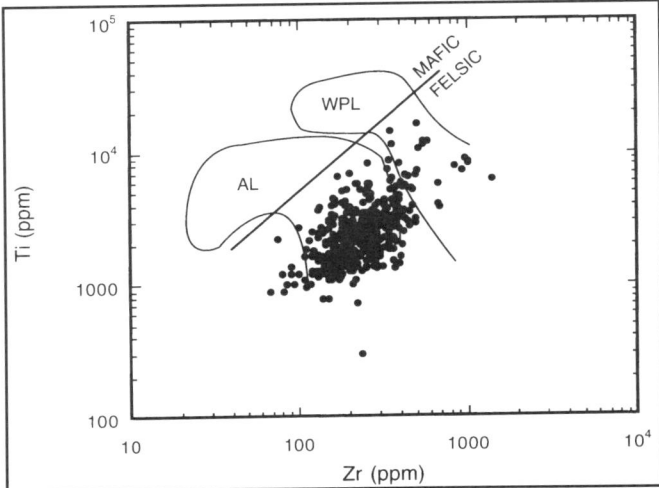

Figure 75. Zr-Ti ratios of Ordovician K-bentonites agree with other discrimination plots when shown on the diagram of Pearce (1982). Parental magmas are felsic, calc-alkaline, and have a dominant arc signature, but one underlain by a substantial volume of continental crust. Abbreviations: arc lava, AL; and within-plate lava, WPL.

EVENT STRATIGRAPHIC IMPLICATIONS

Our studies show that several Ordovician K-bentonite beds are unique stratigraphic markers that can be traced with confidence over wide areas of eastern North America. The individual volcanic eruptions that produced these altered ash beds are unlikely to have lasted more than a few weeks, therefore, the base of each ash layer, for all practical reasons, represents an isochronous surface with time resolution unattainable by other means. The K-bentonite stratigraphy, particularly for the mid-Mohawkian succession, provides detailed information regarding coeval lithofacies, relative age and extent of unconformities, relative rates of sedimentation, timing of basin subsidence and uplift of arches and domes, and paleogeography.

Outcrop and subsurface data were used to interpret the mid-Mohawkian facies relations along a transect from Minnesota to Virginia (Plate 8), a distance of 1,300 km. The most widely traceable mid-Mohawkian K-bentonites include the Hockett, Ocoonita, Deicke, Millbrig, Elkport, and Dickeyville, referred to as the Hagan K-bentonite Complex (Fetzer, 1973; Bergström, 1989). These beds provide the time lines within this tightly controlled stratigraphic succession. The Dickeyville was chosen as the datum on this cross section, because it is widespread and is particularly useful in depicting the stratigraphic relations in the region of the Midcontinent arches and domes. The transect shows the patterns of deposition from the interior craton to the Sevier foreland basin with primary focus on the early to mid-Mohawkian age rocks. The chronostratigraphic framework provided by the K-bentonites reveals detailed information regarding geologic events in the Midcontinent during mid-Mohawkian time:

1. The Middle Ordovician succession in the Upper Mississippi Valley region, situated between the Transcontinental and Wisconsin Arches, consists of a transgressive sequence that includes (in ascending order) the St. Peter Sandstone, Glenwood Shale, and the predominantly carbonate rocks of the Platteville and Galena Groups (Fig. 11). The Decorah Formation, consisting mainly of siliciclastics shed from the Transcontinental Arch, lies at the top of the Platteville pinching-out on the west flank of the Wisconsin Arch. The relative thinness of the K-bentonite–bearing interval (Hagan K-bentonite Complex) in the Platteville and lower Decorah in the Upper Mississippi Valley compared to the southern Appalachians suggests that this is a condensed section. The condensation may extend well up into the Galena Group. Other evidence of a condensed section includes numerous hardgrounds, pellets and encrustations of collophane, locally occurring glauconite grains, and organic matter. Furthermore, the Platteville is characterized by an upward increase in $\delta^{13}C$ that peaked during deposition of the Guttenberg Limestone Member (Hatch et al., 1987). These features indicate that the succession was deposited on the interior platform of North America during a period of maximum transgression in times of low sedimentation rates or nondeposition, long submarine exposure, erosion, and reworking of the sediment in a manner described by Loutit et al. (1988).

2. The Platteville succession thickens southward from 3 m in Freeborn County, Minnesota, to approximately 45 m in St. Charles County, Missouri, to nearly 200 m in the southern part of the Illinois Basin. In northeastern Iowa and southeastern Minnesota the entire Platteville is no older than *P. undatus* Zone (Sweet, 1987), whereas in the Illinois Basin and adjacent regions of Kentucky and Tennessee the carbonate rocks in the basal Platteville are as old as *C. sweeti* Zone (Thomas H. Shaw and Walter C. Sweet, 1994, written communications). Furthermore, the Deicke K-bentonite occurs in a consistent position at the top of the Platteville throughout the eastern Midcontinent region. These stratigraphic relations suggest that the carbonate rocks in the lower three-quarters of the Platteville succession in the Illinois Basin and adjacent regions of Kentucky and Tennessee have no equivalents cratonward in southern Minnesota. The marine transgression accompanying deposition of the Tippecanoe Sequence began during Whiterockian time at the margins of the craton but did not reach the northern Midcontinent until late Turinian time except in the Michigan Basin where the Ibexian-Whiterockian succession appears complete.

3. Distribution of the Hagan K-bentonite complex in the Upper Mississippi Valley clearly demonstrates the time of tectonic uplift on the Wisconsin Arch and the overstepping nature of the Decorah Formation from west to east on the arch (Kolata et al., 1986). During Platteville deposition, yet prior to deposition of the Deicke K-bentonite, the Wisconsin Arch, extending through southern Wisconsin into northern Illinois, uplifted gradually. A general shallowing of the Platteville sedimentary sequence on the arch is suggested by the upward shift from a diverse brachiopod-echinoderm-mollusc–dominated fauna in the Mifflin and Grand Detour Formations (Fig. 11) to less

diverse faunas containing tabulate and rugose corals in the overlying Nachusa and Quimbys Mill Formations (Witzke and Kolata, 1988). Locally, the Quimbys Mill is characterized by unfossiliferous, laminated, mud flat facies clearly indicating depositional shallowing. A prominent unconformity marking a hiatus between the Quimbys Mill and the overlying Decorah is present on the Wisconsin Arch. Furthermore, in a quarry near Argle, Wisconsin (locality 88), significant topographic relief at the top of the Quimbys Mill suggests local subaerial erosion. Traced westward and northwestward through the numerous outcrops in northeastern Iowa and southeastern Minnesota (Kolata et al., 1986) this unconformity appears to correspond to the prominent bedding plane believed to be a diastem at the base of the Carimona Member of the Decorah Formation in southeastern Minnesota (Templeton and Willman, 1963; Mossler, 1985; Kolata et al., 1986). The Carimona thickens west and northwestward from an erosional featheredge at a key section near Dickeyville, Wisconsin (locality 6), to 1.7 m near Rochester, Minnesota (locality 90). The Deicke K-bentonite occurs at or near the base of the Carimona, and the Millbrig is present in the overlying Spechts Ferry Shale Member. A widespread hardground capping the Carimona merges with the unconformity at the top of the Quimbys Mill Formation. The interval between the Deicke and Millbrig, including most of the Carimona and the lower part of the Spechts Ferry, decreases in thickness from 2.7 m at Rochester, Minnesota, to <10 cm near Dickeyville, Wisconsin (locality 6). Progressively younger strata of the Decorah Formation overstep the unconformity from west to east on the Wisconsin Arch. The Millbrig and associated strata extend eastward beyond the erosional featheredge of the Deicke, the Elkport beyond the Millbrig, and the Dickeyville extending even farther to the east. In north-central Illinois and southern Wisconsin, even the Dickeyville and enclosing strata of the Guttenberg Limestone Member pinchout over the arch. As a result, the Dunleith Formation of the Galena Group overlies the Quimbys Mill Formation unconformably locally. We believe that similar overstepping relationships of the Deicke, Millbrig, Elkport, and Dickeyville K-bentonites and associated strata occur in the subsurface along the southern and eastern margins of the Wisconsin Arch. No K-bentonites have been observed in outcrops of the upper Platteville and lower Galena in the vicinity of Appleton, Wisconsin, or in the subsurface at DePere, Wisconsin. This suggests that the Middle Ordovician episode of uplift on the Wisconsin Arch extended over a broad area from Iowa to Michigan, perhaps as much as 250 km wide.

4. The Millbrig is locally absent on the Jessamine Dome in north-central Kentucky and on the Nashville Dome in central Tennessee (Plates 6 and 7), apparently as a result of early Chatfieldian uplift and erosion and/or nondeposition on these structures (Cressman, 1973; Wilson, 1949). Stratigraphic relationships indicate that the main period of uplift occurred after the Deicke ash fall, during or shortly after deposition of the Millbrig. Normal marine sedimentation resumed on these structures shortly before deposition of the Dickeyville K-bentonite. Uplift began somewhat later and ended earlier on the Jessamine and Nashville Domes than on the Wisconsin and Kankakee Arches. Further, uplift of the Nashville Dome is believed to have occurred several times during the late Mohawkian. The distribution of lithofacies and biofacies suggests that a narrow belt of shallow water, the Central Tennessee Bank, was situated along the crest of the dome, and that small islands developed periodically (Wilson, 1962). Thus during early Chatfieldian time, an extensive antiformal system had developed in the eastern Midcontinent. An archipelago consisting of shoals and small islands exposing the exhumed Black River (=Platteville, Tyrone, Carters) carbonates extended from central Wisconsin southward into northern Illinois curving eastward through northern Indiana and southward to central Tennessee (Fig. 76). The setting probably was similar to the present-day Bahama Banks with locally emergent carbonate islands. The archipelago separated the actively subsiding Illinois, Michigan, and Sevier Basins. The Transcontinental Arch and Ozark Dome also underwent uplift at this time, perhaps resulting from the same tectonic forces.

5. The stratigraphic interval of the Hagan K-bentonite Complex is relatively thick within the Illinois Basin, is markedly thin over the Cincinnati Arch, and gradually thickens east of the arch toward the Sevier Basin. Relatively rapid subsidence in the Sevier Basin is particularly evident in the easternmost counties of Kentucky and adjacent parts of western Virginia (Plate 6).

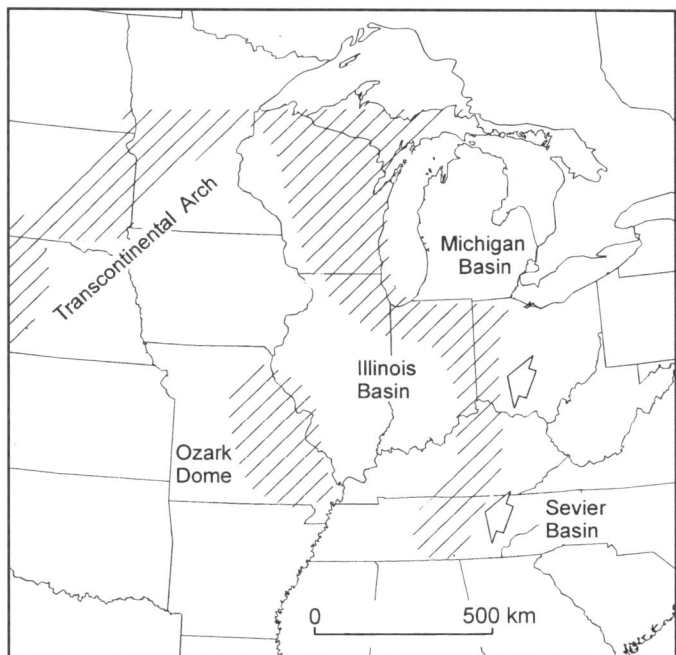

Figure 76. Chatfieldian paleogeography of the eastern Midcontinent showing anticlinal system (hachured) characterized by an archipelago of shoals and small islands extending from central Wisconsin southward into northern Illinois curving eastward through northern Indiana and southward to central Tennessee. The archipelago separated the actively subsiding Illinois, Michigan, and Sevier Basins.

Evidence for a western hingeline of the basin can be seen by comparing the two wireline logs from Johnson County, Kentucky. Within a distance of 20 km, the interval between the Deicke and Millbrig thickens from 8 m to 12 m. Furthermore, the unconformity between the Black River and Trenton Limestones in Kentucky and Tennessee merges with the conformable succession of strata in the Eggleston and Trenton Formations of West Virginia and Virginia and the Stones River and Nashville Groups in northern Alabama (Haynes, 1994).

6. The Midcontinent arches and domes underwent a major episode of uplift concurrent with the volcanism that produced the Hagan K-bentonite Complex. The mechanism that caused the uplifts is not fully understood but it was likely controlled by plate tectonic interactions. A possible explanation is that forces exerted by plate convergence created compressional stresses that propagated across the cratonic interior in a manner suggested by Cloetingh et al. (1985). Recent studies propose that compressive stresses originating from plate collisions during the Paleozoic were transferred to the interior of the craton reactivating long-lived zones of crustal weakness (Kluth, 1986; Kolata and Nelson, 1991; Craddock et al., 1993). This tectonism resulted in basin subsidence, uplift of arches and domes, and reactivation of high-angle faults and associated forced folds (Leighton and Kolata, 1991). Foreland deformation extending as much as 1,300 km away from convergent margins has been documented in the Late Cretaceous collision of Africa and Eurasia and the Miocene collision of India and China (Ziegler, 1988). Accordingly, the following scenario of tectonic events is suggested for eastern Laurentia during mid-Mohawkian time: (1) horizontal compression, caused by the collision of an island arc or microcontinent at the southeastern margin of Laurentia, penetrated deep within the craton causing uplift of many arches and domes while accelerating subsidence in the Illinois and Michigan Basins; (2) subaerial exposure of the uplifted Transcontinental Arch resulted in erosion of igneous terranes and shedding of siliciclastic sediments eastward into the Mississippi Valley region; (3) unconformities developed on the uplifted arches and domes; (4) volcanic eruptions along the active margin ejected large volumes of ash into the atmosphere; (5) the ash was carried by southeast tradewinds more than 1,000 km to the northwest to accumulate in the shallow cratonic seaway; and (6) shortly before deposition of the Dickeyville K-bentonite, most of the arches and domes were quiescent and submerged once again. Realignment of tectonic forces during the remainder of the Taconian orogeny resulted in a change in style and location of deformation. Some structures, such as the Nashville Dome, experienced renewed uplift during late Mohawkian time (Wilson, 1962).

SUMMARY AND CONCLUSIONS

This is the first attempt to establish a comprehensive K-bentonite stratigraphy for the entire Ordovician succession of eastern North America. Although this book is concerned with K-bentonites in eastern North America, we are well aware of coeval beds in Alaska, South America, China, and particularly northern Europe, and we have previously suggested (Huff et al., 1992) the possible equivalence of some beds. Our survey is based on a review of the literature as well as detailed outcrop and subsurface investigations. As pointed out in the Introduction, this book is a progress report that we hope will provide a framework for future investigations. Much additional work is needed before the significance of these unique rocks is fully understood. Major conclusions of this book are summarized below:

1. The Ordovician stratigraphic succession of eastern North America contains at least 60 altered volcanic ash beds, K-bentonites, one or more of which are distributed over an area of 1,500,000 km^2. The beds range in age from Ibexian to Cincinnatian with the greatest concentration in mid-Mohawkian strata. Most beds are not widely distributed but some can be correlated confidently for hundreds of kilometers. Because each ash bed was deposited in a matter of weeks, the base of a K-bentonite, for all practical purposes, is an isochronous surface with time resolution unattainable by other means. The widely distributed K-bentonites, therefore, provide a sound temporal framework useful in determining coeval lithofacies, relative age and extent of unconformities, relative rates of sedimentation, timing of basin subsidence and uplift of arches and domes, and paleogeography.

2. The K-bentonites represent the distal, glass-rich portion of air-fall ash derived from collision-zone explosive volcanism. Overall, the K-bentonites are thicker, coarser, and more numerous toward the central and southern Appalachians, suggesting that the source volcanoes were situated at a latitude between Alabama and Pennsylvania. Major- and trace-element analyses of approximately 750 whole rock K-bentonite samples indicate that the magmatic source was of calc-alkaline composition, mineralogically similar to a suite ranging through andesite, rhyodacite, trachyandesite, and rhyolite. Furthermore, data for 142 K-bentonite samples plotted on Ta-Yb discrimination diagrams (Pearce et al., 1984) show a pattern suggestive of a transition from arc or syn-collision to predominantly within-plate source magmas. The data clearly fall into the category of Th-enriched, calc-alkaline, destructive plate-margin volcanics.

3. Correlation of individual K-bentonites is achieved through a combination of chemical fingerprinting, tracing on wireline logs, and matching of detailed outcrop descriptions, all tied to a biostratigraphic framework based mainly on conodonts and graptolites. The thickest and most widespread beds in eastern North America include the mid-Mohawkian Hockett (new), Ocoonita (new), Deicke, Millbrig, and Dickeyville K-bentonites. These beds can be traced confidently in outcrop and subsurface from the Sevier Basin of western Virginia and eastern Tennessee into the Illinois Basin. The Deicke, Millbrig, and Dickeyville extend even farther northwestward into the Mississippi Valley, a total distance of approximately 1,300 km.

4. The mid-Mohawkian (*B. compressa* Zone) Hockett K-bentonite Bed (new) is present in the upper part of the Hardy Creek Limestone at Hagan, Virginia (locality 59). Its position is

marked by a prominent deflection on the gamma-ray log from the nearby Shell No. 1 L. S. Bales well (locality 164). The bed correlates with R-5 (Miller and Fuller, 1954; Miller and Brosgé, 1954) in the Rose Hill district of southwestern Virginia, "marker bed b" (Stith, 1979, 1986) in the Black River Limestone of southern Ohio, and probably with the Rockvale "metabentonite" (Conkin and Conkin, 1992) in the upper part of the Lebanon Limestone in central Tennessee. The Hockett is present locally in the subsurface of the Illinois and Michigan Basins. Additional stratigraphic studies are needed to determine if the Hockett is equivalent to the locally occurring K-bentonite at the top of the Mifflin Formation in the Upper Mississippi Valley region.

5. The mid-Mohawkian (*B. compressa/P. undatus* Zone) Ocoonita K-bentonite Bed (new) is present in the Eggleston Formation at Hagan, Virginia (locality 59). Its position is also marked by a prominent deflection on the nearby Shell No. 1 L. S. Bales well (locality 164). The Ocoonita correlates with the R-6 (Miller and Fuller 1954; Miller and Brosgé, 1954) in the Rose Hill district of southwestern Virginia, V-2 (Rosenkrans, 1936) in the Moccasin Formation in southwestern Virginia, T-2 (Wilson, 1949) in the Carters Limestone of central Tennessee, B-2 (Fox and Grant, 1944) in the Carters Limestone in southeastern Tennessee, and "marker bed γ" (Stith, 1979, 1986) in the Black River Limestone of southern Ohio. The Ocoonita is locally present in the subsurface of the Illinois and Michigan Basins.

6. The Deicke is the most widespread Ordovician K-bentonite in eastern North America. It can be traced confidently from its type area near St. Louis, Missouri, through nearby outcrops into the subsurface where it produces a prominent and persistent deflection on gamma-ray logs. From the type area, the Deicke is readily traced on wireline logs and drill cores northward to Minnesota and northeastward through the Illinois Basin, to pinch-out on the flanks of the Kankakee Arch in northeastern Illinois and northern Indiana. Northeast of the Kankakee Arch the Deicke is widespread in the Michigan Basin where it is commonly referred to as the "Black River shale." It also appears to be equivalent to K-bentonite MH (Liberty, 1969) in outcrops of the Gull River Formation in southwestern Ontario. Traced on wireline logs and in drill cores and cuttings eastward from St. Louis through the Illinois Basin, it is clear that the Deicke is equivalent to the Pencil Cave K-bentonite of Kentucky, "marker bed β" (Stith, 1979, 1986) in Ohio, T-3 (Wilson, 1949) in Tennessee, R7 (Miller and Fuller, 1954; Miller and Brosgé, 1954) in the western outcrop belt of southwestern Virginia, and V-3 (Rosenkrans, 1936) in the central outcrop belt. The two latter correlations also have been shown by Haynes (1994). Other local synonyms for the Deicke are summarized in Table 1. The Deicke also has been chemically identified in the subsurface as far west as Plymouth County, northwestern Iowa (Kolata et al., 1986, 1987).

7. From its type section in northwestern Illinois, the Millbrig K-bentonite Bed can be traced through the Upper Mississippi Valley outcrop belt and into the subsurface where it is present in drill cores and recorded on wireline logs at numerous localities in eastern Iowa, western Illinois, and eastern Missouri. Correlation of the Millbrig in the Mississippi Valley region has been confirmed also by chemical fingerprinting methods (Kolata et al., 1986, 1987). Wireline correlations of the Millbrig are established along a transect from Missouri northeastward through the Illinois Basin to northeastern Illinois and northern Indiana where the bed pinches-out on the flanks of the Kankakee Arch. The Millbrig is locally present in the Michigan Basin and in the outcrop belt of southwestern Ontario where it appears to correlate with K-bentonite MR described by Liberty (1969). Wireline logs and a small number of drill cores in the Illinois Basin show the Millbrig to be present near the contact between the Black River Group and the overlying Trenton Group. East of the basin the Millbrig clearly correlates with the Mud Cave K-bentonite of Kentucky, "marker bed α" (Stith, 1979, 1986) in Ohio, T-4 (Wilson, 1949) in central Tennessee, R10 (Miller and Fuller, 1954; Miller and Brosgé, 1954) in southwestern Virginia, and according to Haynes (1992, 1994) with V-4 (Rosenkrans, 1936) in the central outcrop belt of Virginia. Other local synonyms for the Millbrig are given in Table 1. Biostratigraphical, geochemical, isotopic, and paleogeographical data suggest that the Millbrig may be equivalent to the Kinnekulle K-bentonite (Bergström et al., 1995) of southern Baltoscandia (Huff et al., 1992).

8. The Dickeyville K-bentonite Bed can be traced from its type section in southwestern Wisconsin through the Upper Mississippi Valley outcrop belt into the subsurface where it is present in drill cores and produces a deflection on some gamma ray wireline logs from northeastern Iowa, western Illinois, and eastern Missouri. The bed appears to correlate with the House Springs K-bentonite (Kolata et al., 1986, 1987) in the outcrop belt of southeastern Missouri; however, this correlation is considered provisional. Like the Deicke and Millbrig, the Dickeyville is readily traced northeastward on wireline logs through the Illinois Basin, to pinch-out on the flanks of the Kankakee Arch in northeastern Illinois and northern Indiana. The Dickeyville does not appear to be very thick nor widespread in the Michigan Basin. Wireline log cross sections transecting the southern part of the Illinois Basin show the Dickeyville to correlate with a prominent and widespread K-bentonite described by Stith (1986) in the basal part of the Lexington Limestone in southern Ohio. Stratigraphic relations suggest that the Dickeyville is equivalent to the Capitol "Metabentonite" described by Conkin and Dasari (1986) in the basal Lexington of northern Kentucky and probably to bed T-5 (Wilson, 1949) in the basal Hermitage Formation of central Tennessee. The Dickeyville is marked by a prominent and persistent deflection on gamma-ray logs in eastern Kentucky and southwestern Virginia and is exposed at Hagan, Virginia (locality 59). Bed V-7 (Rosenkrans, 1936), exposed in the central outcrop belt of southwestern Virginia, is in approximately the same stratigraphic position and may be equivalent to the Dickeyville, but more stratigraphic studies are needed to be certain of this correlation.

9. We suspect that, with additional exploration, several other beds or multiple bed complexes will prove to have event-stratigraphic significance. These include (1) one or two Ibexian (*O. evae* Zone) K-bentonites from the Knox Group in the subsurface of eastern Missouri, southern Illinois, northern Kentucky, eastern Ohio, and northern Alabama; (2) a Whiterockian (*E. reclinatus* Zone) bed in the lower part of the Athens Shale in northeastern Alabama (Drahovzal and Neathery, 1971) and a possible equivalent in the Wells Creek Dolomite of northern Kentucky; (3) an early Mohawkian (*P. aculeata* Zone) bed identified as R1 in the Dot Limestone of southwestern Virginia (Miller and Fuller, 1954), as S1 in the St. Paul Group of West Virginia (Perry, 1964), as an unnamed bed in the Blackford Formation in western Virginia (Heyman, 1970), as B-1 in the Murfreesboro Limestone in southeastern Tennessee (Fox and Grant, 1944), and as an unnamed bed in the Dutchtown Limestone of southeastern Missouri; (4) an early Mohawkian (*B. gerdae* Subzone) bed in the upper Athens Shale in northeastern Alabama (Drahovzal and Neathery, 1971) and possible equivalents, including one of six beds in the Botetourt Limestone Member of the Edinburg Formation in northern Virginia, a bed in the Hurricane Bridge Limestone of southwestern Virginia, and a bed in the Beckett Limestone of southeastern Missouri; (5) several late Mohawkian beds in the Utica Shale (*C. americanus* Zone) of New York (Goldman et al., 1994; Mitchell et al., 1994), several beds in the Montreal and Neuville Formation of the St. Lawrence lowlands (Brun and Chagnon, 1979), the Westboro and Bear Creek K-bentonite zones in the Lexington Limestone and Point Pleasant Formation of southern Ohio (Schumacher and Carlton, 1991), several beds in the Dolly Ridge Formation of West Virginia (Perry, 1972; Ryder, 1992a, 1992b), and the Calmar, Conover, Nasset, and Haldane K-bentonite Beds in the Dunleith Formation of the Upper Mississippi Valley (Willman and Kolata, 1978); (6) the early Cincinnatian Dygerts K-bentonite in the Wise Lake Formation of the upper Mississippi Valley region and possible equivalents in the Kimmswick Limestone of eastern Missouri, a persistent deflection on wireline logs from the Galena Group in the Illinois Basin, a bed in the Sherman Falls Formation in southwestern Ontario (Trevail, 1990), and a bed in the upper part of the Neuville Formation in the St. Lawrence lowlands (Brun and Chagnon, 1979); and (7) a Cincinnatian age (post–*G. pygmaeus* Zone) K-bentonite in the Maquoketa Group in the Seneca County, Ohio, drill core (Bergström and Mitchell, 1992) and its possible equivalents in the Maquoketa Group at Kentland, Indiana (Templeton and Willman, 1963), and one of four beds in the Dubuque Formation in southeastern Minnesota (Levorson et al., 1979).

10. The stratigraphic relations of the Deicke, Millbrig, Elkport, and Dickeyville K-bentonites demonstrate the time of tectonic uplift on many arches and domes in the Midcontinent. These K-bentonites and associated strata overstep the Wisconsin and Kankakee Arches and are eroded from the Jessamine and Nashville Domes, showing that the episodes of uplift took place at approximately the same time the ash beds were deposited. An extensive anticlinal system developed in the Midcontinent during the late Turinian. An archipelago of shoals and small islands exposing the exhumed Black River (=Platteville, Tyrone, Carters) carbonates extended from central Wisconsin southward into northern Illinois, curved eastward through northern Indiana and then back southward through northern Kentucky into central Tennessee. The archipelago separated the actively subsiding Illinois, Michigan, and Sevier Basins.

11. During mid-Mohawkian time, collision of a microcontinent at the southeastern margin of Laurentia caused horizontal compressive stress to be transmitted deep within the craton reactivating long-lived zones of crustal weakness. This resulted in uplift of many arches and domes and perhaps accelerated subsidence within the basins. Large volumes of ash ejected from volcanoes situated along the southeastern margin of Laurentia were carried by the prevailing southeast tradewinds for hundreds of kilometers northwestward to accumulate in shallow cratonic seas. During the remainder of the Taconian orogeny, compressional stresses realigned and the locus of tectonic activity, including volcanism, shifted northward along the margin of Laurentia. Most of the Midcontinent arches and domes remained relatively stable for the rest of the Ordovician.

ACKNOWLEDGMENTS

We thank Raymond L. Ethington, Richard A. Hoppin, Janis D. Treworgy, and Malcolm P. Weiss for the many helpful comments and suggestions on an early draft of this book. We also thank Curtis Abert for help in plotting the data in the Geographic Information System and Jacquelyn L. Hannah for assisting in producing the computer-generated illustrations and wireline cross sections. For assistance in acquiring subsurface and outcrop data, we thank Derek K. Armstrong, Katharine Lee Avary, Bill J. Bunker, Marvin P. Carlson, Terry Carter, Richard J. Diecchio, John Doveton, James A. Drahovzal, Ronald C. Elowski, William B. Harrison III, John T. Haynes, Stephen A. Leslie, Greg A. Ludvigson, Charles E. Mitchell, John H. Mossler, Jack E. Nolde, Eugene K. Rader, Ronald A. Riley, Ira R. Satterfield, Gregory A. Schumacher, David A. Stith, Paul Sundeen, Thomas L. Thompson, Robert A. Trevail, Lawrence H. Wickstrom, and Brian J. Witzke. Unpublished Middle Ordovician conodont data provided by Thomas H. Shaw and Walter C. Sweet for the Upper Mississippi Valley and Illinois Basin regions were of particular help. This research was supported by National ScienceFoundation grants EAR-9004559 (SMB), EAR-9205981 (SMB), EAR-8025889 (WDH, DRK), EAR-8208480 (WDK, DRK), EAR-8407018 (WDH, DRK), EAR-8904295 (WDH, DRK), EAR-9005333 (WDH, DRK), and EAR-9204893 (WDH, DRK).

APPENDIX 1. LOCALITY REGISTER

1. Quarry in lower part of the St. Peter Sandstone 0.8 km south of Hanover, Rock County, Wisconsin (SE¼ SE¼ SW¼ sec. 14, T. 2 N., R. 11 E.). Slabs of feldspathized K-bentonite dredged from the pit are found in the surrounding talus slopes.

2. Ravine exposure 5 km west of Utica, LaSalle County (NW¼ NE¼ sec. 13, T. 33 N., R. 1 E.). Probable K-bentonite beds at base of the St. Peter Sandstone (Cady, 1919).

3. Cement company quarry north of State Highway 2 in Dixon, Lee County (sec. 27, T. 22 N., R. 9 E.). Medusa Section 61 (Willman and Kolata, 1978).

4. Railroad cut along south bluff of Meramec River along the St. Louis–San Francisco Railroad, 0.4 km southwest of Tyson, St. Louis County, Missouri (near center E½ SE¼ SE¼ sec. 21, T. 44 N., R. 4 E.). Type section of the Deicke K-bentonite (Willman and Kolata, 1978). Mincke Hollow Section 60 (Kolata et al., 1986).

5. Natural outcrop along east bank of Galena River, 1.6 km southeast of Millbrig, Jo Daviess County, Illinois (near sec. 34, T. 29 N., R. 1 E.). Type section of the Millbrig K-bentonite (Willman and Kolata, 1978). Millbrig Southeast Section 45 (Kolata et al., 1986).

6. Quarry and roadcut on U.S. Highway 61, 6.4 km northwest of Dickeyville, Grant County, Wisconsin (NE¼ NE¼ SW¼ sec. 7 and S½ NW¼ sec. 7, T. 2 N., R. 2 W.). Dickeyville Northwest Section 40 (Kolata et al., 1986). Deicke, Millbrig, Elkport, and Dickeyville K-bentonites (type section of the latter).

7. Quarry on east side of State Highway 340, 1.2 km south of intersection of highways 340 and 18 in McGregor, Clayton County, Iowa (SW¼ SW¼ SE¼ sec. 27, T. 95 N., R. 3 W.). McGregor South Section 23 (Kolata et al., 1986). Deicke, Millbrig, and Elkport K-bentonite Beds.

8. Quarry along the Burlington Northern Railroad 1.6 km west of Polo, Ogle County, Illinois (SE¼ NE¼ NE¼ sec. 18, T. 23 N., R. 8 E.). Polo West Section (Kolata et al., 1986). Dickeyville K-bentonite.

9. Kapper Fountain Quarry, 0.8 km west of Fountain, Fillmore County, Minnesota (SW¼ SW¼ sec. 3, T. 103 N., R. 11 W.). Unnamed K-bentonite 2 cm thick near middle of the Fairplay Limestone Member of the Dunleith Formation, 2.6 m above floor of quarry. Unpublished section M-102 (Calvin O. Levorson and Arthur J. Gerk, 1972).

10. Pavlovec Quarry, west side of Decorah, Winneshiek County, Iowa (SW¼ NE¼ sec. 18, T. 98 N., R. 8 W.). Pavlovec West Quarry (21-A section, Calvin O. Levorson and Arthur J. Gerk, 1972). Calmar K-bentonite 3.5 m above base of Rivoli Limestone Member of the Dunleith Formation; Conover K-bentonite at top of Rivoli Limestone; Nasset K-bentonite 1.8 m above base of Sherwood Limestone Member of Dunleith Formation.

11. Roadcut of U.S. Highway 52 where it ascends to the upland from the Mississippi River bottomland on the northwest side of Guttenberg, Clayton County, Iowa (SW¼ SW¼ sec. 5, T. 92 N., R. 2 W.). Unnamed K-bentonite 2 cm thick near middle of the Fairplay Limestone Member of the Dunleith Formation 3.4 m above the Eagle Point Limestone Member. Nasset K-bentonite Bed, 4 cm thick, exposed 30 m above base of outcrop. Guttenberg North Section (Templeton and Willman, 1963). Type section for the Elkport K-bentonite Bed (Willman and Kolata, 1978).

12. Boyles Quarry on east side of State Highway 340, 1.2 km south of intersection of highways 340 and 18 in McGregor, Clayton County, Iowa (SW¼ SW¼ SE¼ sec. 27, T. 95 N., R. 3 W.). Deicke, Millbrig, and Elkport K-bentonite Beds exposed. McGregor South Section 23 (Kolata et al., 1986).

13. Pederson Quarry, 0.8 km north and 2.4 km west of Harmony, Fillmore County, Minnesota (NE¼ SW¼, sec. 9, T. 101 N., R. 10 W.). Calmar K-bentonite Bed 2 cm thick near middle of the Rivoli Limestone Member 5.2 m above quarry floor. Unpublished section M-107 (Calvin O. Levorson and Arthur J. Gerk, 1973).

14. Roadcut on north side of Iowa State Highway 9, 4.3 km east of junction of Highway 9 and U.S. Highway 52 on south side of Decorah, Winneshiek County, Iowa (SW¼ SE¼ sec. 23, T. 98 N., R. 8 W.). Calmar K-bentonite Bed up to 6 cm thick near middle of the Rivoli Limestone Member approximately 4.3 m above road level. Unpublished section 8-C (Calvin O. Levorson and Arthur J. Gerk, 1971).

15. Quarry and diversion channel at Decorah, Winneshiek County, Iowa (NW¼ NE¼ sec. 5, T. 98 N., R. 8 W.). Calmar K-bentonite up to 6 cm thick near middle of the Rivoli Limestone Member 23.5 m above base of outcrop. Outcrop is adjacent to the type section of the Calmar (SE¼ NE¼ sec. 18, T. 98 N., R. 8 W.; Willman and Kolata, 1978).

16. Excavation for diversion channel on west side of Decorah, Winneshiek County, Iowa (SE¼ NE¼ sec. 18, T. 98 N., R. 8 W.). Type section of the Calmar and Conover K-bentonite Beds (Willman and Kolata, 1978).

17. Roadcut on County Road X-26, 2.4 km northeast of Volney, Allamakee County, Iowa (SW¼ NE¼ sec. 12, T. 96 N., R. 5 W.). Conover K-bentonite Bed, 6 cm thick, exposed 2 m below top of bedrock. Unpublished section 38 (Calvin O. Levorson and Arthur J. Gerk, 1972).

18. Quarry situated 2.8 km east of Wykoff, Fillmore County, Minnesota (NE¼ SW¼ sec. 25, T. 103 N., R. 12 W.). Nasset K-bentonite Bed exposed 6.7 m above floor of quarry. Unpublished section M-101 (Calvin O. Levorson and Arthur J. Gerk, 1973).

19. Ludlow Quarry situated 11 km southwest of Waukon, Allamakee County, Iowa (SE¼ SE¼ sec. 34, T. 95 N., R. 6 W.). Calmar, Conover, and Nasset K-bentonite Beds at 6.1 m, 7.3 m, and 9.6 m above quarry floor, respectively. Unpublished section 39-A (Calvin O. Levorson and Arthur J. Gerk, 1972).

20. Quarry on west side of Pine Creek north of Victory School, 3.2 km southwest of Mt. Morris, Ogle County, Illinois (NW¼ SW¼ SE¼ sec. 32, T. 24 N., R. 9 E.). Type section for the Haldane K-Bentonite Bed. Section 47 (Willman and Kolata, 1978).

21. Quarve & Anderson Quarry situated on east side of U.S. Highway 63, 6.2 km north of junction of U.S. 63 and Interstate 90, 3.2 km south of Rochester, Olmstead County, Minnesota (SW¼ NE¼ sec. 35, T. 106 N., R. 14 W.). Haldane K-bentonite Bed situated in middle of Wall Limestone Member of the Dunleith Formation 11.6 m above quarry floor. Dygerts K-bentonite Bed exposed in roadcut on U.S. 63, 300 m west of the quarry, 7 m above road level. Unpublished section M-115 (Calvin O. Levorson and Arthur J. Gerk, 1973).

22. Quarry on west side of Illinois Central Gulf Railroad on south side of Dixon, Lee County, Illinois (SW¼ SE¼ NE¼ sec. 8, T. 21 N., R. 9

E.). Haldane K-bentonite Bed situated 3.7 m below top of bedrock. Dixon South Section 59 (Willman and Kolata, 1978).

23. Roadcut on U.S. Highway 20 on the east side of the Galena River in Galena, Jo Daviess County, Illinois (SW¼ SE¼ NE¼ sec. 20, T. 28 N., R. 1 E.). Type section of the Dygerts K-bentonite Bed which occurs in the Sinsinawa Dolomite Member of Wise Lake Formation 7.3 m above road level. Geologic section 33 (Templeton and Willman, 1963) and section 14 (Willman and Kolata, 1978).

24. Ingolf Hovey Quarry, 2.4 km south of Decorah, Winneshiek County, Iowa (NW¼ SW¼ sec. 28, T. 98 N., R. 8 W.). Dygerts K-bentonite Bed in the Sinsinawa Limestone Member of the Wise Lake Formation 2 m above quarry floor. Unpublished section 13 (Calvin O. Levorson and Arthur J. Gerk, 1971).

25. Breuning Skyline Quarry, 0.8 km northeast of Decorah, Winneshiek County, Iowa (NW¼ SE¼ sec. 10, T. 98 N., R. 8 W.). Dygerts K-bentonite Bed in Sinsinawa Limestone Member of the Wise Lake Formation 7.5 m above quarry floor. Unpublished section 14 (Calvin O. Levorson and Arthur J. Gerk, 1971).

26. Rockford Black Top Quarry–Nimtz Road, 1.6 km southeast of Harlem, Winnebago County, Illinois (N center SE¼ sec. 33, T. 45 N., R. 2 E.). Dygerts K-bentonite Bed in Sinsinawa Dolomite Member of the Wise Lake Formation 6 m below top of bedrock. Harlem Southeast Section (Willman and Kolata, 1978).

27. Mystery Cave, Spring Valley, Fillmore County, Minnesota (SE¼ sec. 19, T. 102 N., R. 12 W.). Exposure of two unnamed feldspathized K-bentonites in the Dubuque Formation 1 m and 6 m below the contact with the overlying Maquoketa Group. Section described by Weiss (1954) and Levorson et al. (1979).

28. Kapper Construction Company Quarry, 7.4 km north of Spring Valley, Fillmore County, Minnesota (NW¼ SW¼ sec. 3, T. 103 N., R. 13 W.). Unpublished section M-118 (Calvin O. Levorson and Arthur J. Gerk, 1973). Four unnamed K-bentonite beds in the Dubuque Formation occur at 18 m, 19.5 m, 21 m, and 23.5 m above quarry floor.

29. Quarry 2.4 km southeast of Kendallville, Winneshiek County, Iowa (W½ sec. 2, T. 99 N., R. 10 W.). K-bentonite 7.5 cm thick in upper part of the Dubuque Formation, 0.5 m below the Maquoketa Group. Unpublished section 60 (Calvin O. Levorson and Arthur J. Gerk, 1973).

30. Rifle Hill Quarry, 1.6 km north and 4 km east of Cherry Grove, Fillmore County, Minnesota (NE¼ NW¼ sec. 35, T. 102 N., R. 12 W.). K-bentonite 10 cm thick occurs in the Maquoketa Group 5 m above contact with the Dubuque Formation. Unpublished section M-106 (Calvin O. Levorson and Arthur J. Gerk, 1973).

31. Wubbels' Ravine, 3.2 km west of Greenleafton, Fillmore County, Minnesota (N½ sec. 2, T. 101 N., R. 12 W.). K-bentonite 6 cm thick occurs in the Maquoketa Group 6 m above contact with the Dubuque Formation. Section 168 (Weiss, 1957).

32. Roadcut along I-55, 1.3 km south of intersection of I-55 and State Route Z, Pevely, Jefferson County, Missouri (SW NW sec. 19, T. 41 N., R. 6 E., Herculaneum 7.5 minute quadrangle). Unnamed K-bentonite in Beckett Limestone.

33. Roadcut along U.S. Route 61, 8 km west of Ste. Genevieve, Ste. Genevieve County, Missouri (NE SE sec. 28, T. 38 N., R. 8 E., Weingarten 7.5 minute quadrangle). Section described by Thompson (1991). Three unnamed K-bentonites in the Macy Limestone.

34. Roadcut along I-55, 1.5 km north of intersection of I-55 and State Route Z, Pevely, Jefferson County, Missouri (SE sec. 12, T. 41 N., R. 5 E., Herculaneum 7.5 minute quadrangle). Unnamed K-bentonite in the Hook Limestone Member of the Macy Limestone.

35. Roadcut on north frontage road of I-44, west of the Allenton–Six Flags amusement park exit, in southern St. Louis County (SW¼ SW¼ sec. 33, T. 44 N., R. 3 E., Eureka 7.5 minute quadrangle). Millbrig and Deicke K-bentonite Beds.

36. Exposure in Apple Creek tributary 2 km south of Altenburg, Perry County, Missouri (SW¼ SE¼ sec. 28, T. 34 N., R. 13 E., Neelys Landing 7.5 minute quadrangle). Millbrig and House Springs K-bentonite Beds.

37. Roadcut in southeast bluff of Establishment Creek where creek bends abruptly south then back north, 5 km northeast of Bloomsdale, Ste. Genevieve County, Missouri (unnumbered section 0.5 km northwest of sec. 5, T. 38 N., R. 8 E.). Bloomsdale Section 64 in Kolata et al. (1986). Deicke and Millbrig K-bentonites, two unnamed K-bentonites in the Macy Limestone, two feldspathized K-bentonites in the Decorah Formation and the Dickeyville (House Springs K-bentonite) in the base of the Kimmswick Limestone.

38. Roadcut on U.S. Highway 61, 6.4 km southeast of New London, Ralls County, Missouri (SE¼ NE¼ SE¼ sec. 21, T. 55 N., R. 4 W., Hannibal 15 minute quadrangle). Section figured in Kolata et al. (1986). Deicke and Millbrig K-bentonites.

39. Quarry on Klein farm on southwest side of Interstate Highway I-55, 3.2 km southeast of intersection of I-55 and State Highway 32, 7 km southwest of Ste. Genevieve, Ste. Genevieve County, Missouri (southern part of irregular sec. 7, T. 37 N., R. 9 E., Ste. Genevieve 7.5 minute quadrangle). Section 65 in Kolata et al. (1986). Millbrig K-bentonite exposed in roadside ditch leading into quarry.

40. Grays Point Quarry 3 km northeast of Illmo, Scott County, Missouri (390 m east and 120 m north of the northeast corner of sec. 28, T. 30 N., R. 14 E., Thebes 7.5 minute quadrangle). Section 66 in Kolata et al. (1986). Deicke K-bentonite exposed in high wall.

41. Dundee Cement Company quarry 4 km west of Clarksville, Pike County, Missouri (SE¼ sec. 12, T. 53 N., R. 1 W.). Unnamed K-bentonite 5 cm thick in upper part of the Kimmswick Limestone, 20 m below the Maquoketa Group.

42. Natural exposure of along south bank of Calumet Creek, 2.4 km west of Clarksville, Pike County, Missouri (NE¼ SE¼ sec. 18, T. 53 N., R. 1 E.). Unnamed K-bentonite in Maquoketa Group 3.4 m above level of Calumet Creek.

43. Means Quarry, 4 km east of Kentland, Newton County, Indiana (NW¼ NE¼ NW¼ sec. 25, T. 27 N., R 9 W.). Two K-bentonite beds in the Scales Formation of the Maquoketa Group approximately 17 m above the base of the Maquoketa (Templeton and Willman, 1963).

44. Lexington Limestone Quarry, 1.5 km west of U.S. Highway 27 on Catnip Hill Road at Nicholasville, Jessamine County, Kentucky. Deicke K-bentonite. Section 33 (Haynes, 1994).

45. Black River Mine, near Carntown, Pendleton County, Kentucky (5-BB-64). Deicke and Millbrig K-bentonites.

46. Quarry and cut along Southern (CNO&TP) Railroad at High Bridge, Mercer County, Kentucky. Deicke and Millbrig K-bentonites.

47. Outcrop on U.S. Highway 68, 1.6 km southwest of intersection with State Route 33, at entrance to Shakertown, Mercer County, Kentucky (Harrodsburg 7.5 minute quadrangle). Millbrig K-bentonite exposed in roadcut. Deicke can be dug-out in small ditch at south end of roadcut.

48. Outcrop on west side of State Highway 53, 4 km south of intersection with U.S. Highway 70N at South Carthage, Smith County, Tennessee (Gordonsville 7.5 minute quadrangle). Deicke is well exposed in a high wall of the Carters Limestone. The Millbrig, however, is covered by grass at the top of the outcrop. The Millbrig is also exposed at the top of a roadcut at the intersection of State Highway 53 and U.S. Highway 70N.

49. Tillet Brothers Paving Company Quarry (the "Stone Man") south of U.S. Highway 41 Alternate, 8 km southeast of Shelbyville, Bedford County, Tennessee (Normandy 7.5 minute quadrangle). Deicke and Millbrig exposed in quarry highwall.

50. Davis Crossroads, Walker County, Georgia. Cut along the abandoned Southern Railroad ~800 m southeast of the grade crossing at State Road 341. Exposure of the Deicke and Millbrig K-bentonite beds in the Carters Limestone. Section described by Allen and Lester (1957), Milici and Smith (1969) and Haynes (1994).

51. Rising Fawn, Dade County, Georgia. Cut along the Southern Railway ~2 km south of the town of Rising Fawn. Exposure of the Deicke and Millbrig K-bentonite Beds in the upper Carters Limestone. Section described by Haynes (1994).

52. Fort Payne, DeKalb County, Alabama. Roadcut along southeast side of the northbound lane of Interstate Highway 59 just north of Interchange 222. Exposure of the Deicke and Millbrig and several other K-bentonites in the Stones River Group. Section described by Drahovzal and Neathery (1971), Huff and Kolata (1990) and Haynes (1994).

53. Roadcut along expressway cut through Red Mountain at Birmingham, Alabama, Jefferson County (central 1/2, sec. 6, T. 18 S., R. 2 W., Birmingham South 7.5 minute Quadrangle). See Drahovzal and Neathery (1971) for detailed stratigraphic section. Section 29 of Haynes (1994).

54. Southern Cement Shale Quarry on west side of narrow south-southwest–trending ridge near Calera, Shelby County, Alabama (NW1/4 NE1/4 sec. 5, T. 24 N., R. 13 E., Montevallo 7.5 minute Quadrangle). Unnamed K-bentonite in lower part of the Athens Shale. See Drahovzal and Neathery (1971) for stratigraphic description.

55. Old North Ragland Quarry, southwest end of short ridge near Ragland, St. Clair County, Alabama (NE1/4, sec. 20, T. 15 S., R. 5 E., Ragland 7.5 minute quadrangle). Two unnamed K-bentonites in the Little Oak Limestone.

56. Outcrop of Athens Shale in Page Springs Anticline, 3 km north of Wilsonville, Shelby County, Alabama (NE1/4 SW1/4 NW1/4 sec. 19, T. 20 S., R. 1 E. Locality 5 of Drahovzal and Neathery (1971). Unnamed K-bentonite in upper part of the Athens Shale.

57. Greensport Gap, Etowah County, Alabama; locality from Neathery and Drahovzal (1985), section measured along Alabama Highway 77, at gap through Greens Creek Mountain in (sec. 6, T. 14 S., R. 6 E., beginning in SE1/4 NW1/4 sec. 6 and ending approximately 30 m south of unmarked county road in SE1/4 SE1/4 SW1/4 sec. 6, Ohatchee 7 1/2 minute Quadrangle) exposure of Colvin Mountain Sandstone containing two K-bentonites, one of which could be the Millbrig.

58. Alexander Gap, Calhoun County, Alabama; locality of Haynes (1994); Colvin Mountain Sandstone containing the Deicke and Millbrig K-bentonite Beds exposed along east side of northbound U.S. Highway 431 in Alexander Gap just south of the initial downgrade through the gap.

59. Cut on the east side of the Louisville & Nashville Railroad tracks at Hagan, Lee County, Virginia. Section described by Huffman (1945) and Miller and Fuller (1954). Several K-bentonites occur in the Eggleston Formation including the Deicke and Millbrig. Type section of the Ocoonita and Hockett K-bentonites (new).

60. Tazewell, Tazewell County, Virginia; Moccasin, Eggleston, and Trenton Formations containing 13 K-bentonite beds exposed 800 m north of the intersection of Virginia State Highway 16 and Tazewell County Road 604 near Tazewell, Tazewell County, Virginia. Section described by Rosenkrans (1936) and Fetzer (1973).

61. Road cut at south end of Southern Railroad bridge on State Highway 348, 1.4 km south-southwest of Mosheim, Greene County, Tennessee. Lower part of the Blockhouse Shale containing five unnamed K-bentonite beds.

62. Catawba, Roanoke County, Virginia; upper part of Bays Formation in exposure along southwest side of Route 311, 50 m southeast of bridge over Catawba Creek; five K-bentonites exposed including the Millbrig and V-7; Catawba 7 1/2 minute quadrangle.

63. Tumbling Run Section—natural outcrops along Tumbling Run 0.48 km southwest of junction on U.S. Highway 11 and State Route 601 on southwest side of Strasburg, Shenandoah County, Virginia. Six K-bentonite beds as much as 25 cm thick are present in the basal part of the Edinburg Formation. Section described in Cooper and Cooper (1946) and Rader and Read (1989).

64. Oranda Type Section—outcrop of the Middle Ordovician Edinburg, Oranda, and Martinsburg Formations on State Highway 55, 0.64 km northwest of intersection with U.S. Highway 11 in Strasburg, Shenandoah County, Virginia. Measured section (Cooper and Cooper, 1946). Fieldtrip stop 2 of Rader and Read (1989).

65. Colliertown Section—outcrops along State Highway 251 3.2 km east of Colliertown, Rockbridge County, Virginia. Geologic section 32 (Cooper and Cooper, 1946). K-bentonite bed 4 cm thick near middle of the Edinburg Formation.

66. Mauzy Section—outcrop on northeast side of State Route 608, 0.5 km southeast of the intersection of Route 608 and U.S. Highway 11 in Mauzy, Rockingham County, Virginia. Geologic section 5 (McVey, 1993). Outcrop of the upper part of the Oranda Formation and lower part of the Martinsburg Formation containing a succession of K-bentonite beds including the Deicke and Millbrig.

67. Martinsburg Section—Capitol Cement Corporation quarry off Route 9 on the south side of Martinsburg, Berkeley County, West Virginia. Geologic section 7 of McVey (1993). Limestone of the Chambersburg Formation overlain by the Martinsburg Shale containing 12 K-bentonite beds including the Deicke and Millbrig. 78° 10' 50" longitude, 40° 37' 10" latitude.

68. Union Furnace—outcrop of the Clover, Milroy, Snyder, Linden Hall, Nealmont, Salona, and Coburn Formations in the Warner Company Quarry ~1.2 km north of Union Furnace, Huntington County, Pennsylvania. The New Enterprise Quarry is situated adjacent to the Union Furnace Quarry. Section described by Kay (1944), Thompson

(1963), Rones (1969), and Smith et al. (1986). Section 8 (McVey, 1993) given as 78° 10′ 50″ west longitude and 40° 37′ 10″ north latitude.

69. Oak Hall—outcrop of the Clover through Salona Formations in the Neidigh Brothers quarry 0.8 km northwest of Oak Hall, Centre County, Pennsylvania. Section described by Thompson (1963) and Rones (1969).

70. Pleasant Gap—outcrop of the Snyder through Salona Formations in the Whiterock quarries at Pleasant Gap, Centre County, Pennsylvania. Section described by Rones (1969).

71. Roadcut on State Highway 12, 3.2 km north of Boonville, Oneida County, New York. Four K-bentonites exposed in the Napanee and Kings Falls Limestones.

72. Hounsfield type section—quarry 3.2 km east of Dexter, Jefferson County, New York. Section described by Kay (1931).

73. Middleville Section—quarry on east side of County Route 169 ~1.2 km southeast of Middleville, Herkimer County, New York.

74. Ravine exposure along Buttermilk Creek ~4.8 km north of Middleville, Herkimer County, New York.

75. Coldwater Quarry at Coldwater, Simcoe County, Ontario. K-bentonites MX and MH occur in outcrop (Liberty, 1969).

76. Roadcut on Provincial Highway 7 on west side of Marmora, Hastings County, Ontario. Long exposure of the Gull River Formation containing K-bentonites MX and MN. Marmora West Composite Section (Liberty, 1969).

77. Roadcut on Provincial Highway 68, 6.4 km south of Whitefish Falls, Ontario, on Manitoulin Island (Coordinates 403:011; 41 I/4). K-bentonite (Deicke) exposed near the top of the Swift Current Formation (=Gull River Formation).

78. Roadcut on Provincial Highway 68, 12 km south of Whitefish Falls (coordinates 374.955; 41 I/4). 1.5 cm K-bentonite exposed 15 cm below contact between the Swift Current/Cloche Island Formations.

79. Natural outcrops along the banks of the Escanaba River at Bony Falls, Delta County, Michigan (sec. 1, T. 41. N., R. 24 W.). Two unnamed K-bentonites occur in the upper 4 m of the Bony Falls beds (Black River Group). Section described by Hussey (1952).

80. Outcrops along the banks of the Escanaba River north and south of the concrete bridge 0.8 km northeast of Cornell, Delta County, Michigan (cen. N^1/$_2$ sec. 32, T. 41 N., R. 23 W.). Section described by Templeton and Willman (1963). Unnamed K-bentonite in Trenton Group.

81. Outcrops along the west bank of the Escanaba River below the dam at Chandler Falls, 9.6 km north Escanaba, Delta County, Michigan (NE^1/$_4$ SW^1/$_4$ sec. 25, T. 40 N., R. 23 W.). Section described by Templeton and Willman (1963). Two unnamed K-bentonites in the Trenton Group (Sherman Falls Formation in Kay, 1935).

82. Constructions Deschênes Ltée Quarry, 10 km southwest of Hull, Quebec, Canada, near Aylmer. Section 1 (Brun and Chagnon, 1979). Probable Deicke K-bentonite.

83. Turnbull Construction Inc. Quarry, 5 km northeast of Joliette, Quebec, Canada. Section 12 (Brun and Chagnon, 1979). Probable Deicke K-bentonite.

84. Quarry situated on the south side of the highway in West Bay Centre, 1.75 km due west of South Head, Port-au-Port, western Newfoundland. Three K-bentonites observed in the Table Head Formation by Finney and Skevington (1979).

85. Natural outcrop of Cape Cormorant Formation along Caribou Brook and in steeply dipping beds which form cliffs along the coast northeast of the brook, 0.5 km south of the fishermen's landing in the village of Mainland, western Newfoundland (James et al., 1988).

86. Murray Lane Section, west side of U.S. Highway 99, 5 km south of Fittstown, Pontotoc County, Oklahoma (NW^1/$_4$ SW^1/$_4$ sec. 12, T. 1 N., R. 6 E). Section 8a described by Decker (1933) and section 11A by Fay and Graffham (1982).

87. Rock Crossing Section, Criner Hills, south of Ardmore, Carter County, Oklahoma (sec. 35, T. 5 S., R. 1 E.). Exposure of Viola Limestone exposed below the dam on the northwest side of the bridge over Hickory Creek. Section 10 described by Decker (1933). Possible K-bentonite described as a 3 to 4 inch residual clay bed separating the base of the Viola from the Bromide Formation.

88. Quarry on north side of county road 7 km southwest of Argyle, Lafayette County, Wisconsin (SE^1/$_4$ NW^1/$_4$ SW^1/$_4$ sec. 6, T. 2 N., R. 5 E.). Apparent subaerial erosion at top of Quimbys Mill Formation filled in with strata of the Guttenberg Limestone Member.

89. Outcrop in bluff of Mississippi River at end of Summit Avenue in St. Paul, Ramsey County, Minnesota (E^1/$_2$ NE^1/$_4$ SW^1/$_4$ sec. 5, T. 28 N., R. 23 W.). St. Paul Section (Kolata et al., 1986). Deicke and Millbrig K-bentonites.

90. Quarry on old U.S. Highway 52 on southeast side of Rochester, Olmsted County, Minnesota (SW^1/$_4$ SE^1/$_4$ NE^1/$_4$ sec. 21, T. 106 N., R. 13 W.). Rochester Southeast Section of Kolata et al. (1986). Deicke and Millbrig K-bentonites.

91. Quarry on north side of county road, 0.4 km west of Minnesota State Route 43 and 2 km north of Mabel, Fillmore County, Minnesota (SE^1/$_4$ SW^1/$_4$ SE^1/$_4$ sec. 15, T. 101 N., R. 8 W.). Mabel Northwest Section of Kolata et al. (1986). Deicke and Millbrig K-bentonites.

92. Quarry on east side of county road, 2 km east of Hanover, Allamakee County, Iowa (NW^1/$_4$ NW^1/$_4$ SE^1/$_4$ sec. 25, T. 99 N., R. 6 W.). Hanover East Section (Kolata et al., 1986). Deicke and Millbrig K-bentonites.

93. Quarry 0.4 km south of State Highway 75, 0.8 km east of Rock City, Stephenson County, Illinois (SE^1/$_4$ SW^1/$_4$ NW^1/$_4$ sec. 22, T. 28 N., R. 9 E.). Rock City East Section (Kolata et al., 1986). Dickeyville K-bentonite.

94. Quarry on east side of U.S. Highway 51 in Rochelle, Ogle County, Illinois (NE^1/$_4$ NE^1/$_4$ SW^1/$_4$ sec. 13, T. 40 N., R. 1 E.).

95. Cominco No. SS-8 Welter, Jones County, Iowa (SW^1/$_4$ SW^1/$_4$ NE^1/$_4$ sec 7, T. 85 N., R. 1 W.). Iowa Geological Survey Bureau core no. 27580.

96. A. Jordan No. 1 (W-13121), Columbus Junction, Louisa County, Iowa (NW^1/$_4$ SW^1/$_4$ sec. 32, T. 75 N., R. 4 W.). Dickeyville K-bentonite Bed in upper part of Guttenberg Formation at depth of 283.2 m.

97. C. E. Brehm and Duke Res. No. 1 Clayton Hrs., Franklin County,

Illinois (990 ft from the south line, 330 ft from the west line of SE^1/$_4$ NE^1/$_4$ sec. 2, T. 5 S., R. 3 E.). K-bentonite bed observed by Howard R. Schwalb in drill cuttings at a depth of 7,250 ft in the Shakopee Formation.

98. Texas Pacific Oil No. 1 Mary Streich well in Pope County, Illinois (1,815 ft from the south line and 330 ft from the west line of SE^1/$_4$ sec. 2, T. 11 S., R. 6 E.). K-bentonite bed observed by Howard R. Schwalb in drill cuttings at a depth of 8,260 ft in the upper part of the Shakopee Formation. The apparent petrophysical response to the K-bentonite occurs at a depth of 8,265 ft (2519 m).

99. Conoco, Inc. No. 1 Einar Dyhrkopp, Gallatin County, Illinois (2,300 ft from the north line and 1,650 ft from the east line of sec. 4, T. 10 S., R. 9 E.).

100. Humble No. 1 Pickle, Union County, Illinois (NW^1/$_4$ SE^1/$_4$ NW^1/$_4$ sec. 21, T. 13 S., R. 2 W.). K-bentonite bed observed by Howard R. Schwalb in drill cuttings at a depth of 2,090 ft in the Shakopee Formation. The apparent petrophysical response to a K-bentonite bed occurs at a depth of 2,111 ft on the wireline log.

101. Arthur Job No. 1, Cape Girardeau County, Missouri (NW^1/$_4$ NW^1/$_4$ sec. 1, T. 30 N., R. 13 E.). Sample studies on file at the Missouri Geological Survey report a K-bentonite containing biotite at 110 ft to 120 ft (33.5 m to 36.5 m) below the top of the Dutchtown Limestone.

102. Orchard No. 1 Theo. Ochs, Cape Girardeau County, Missouri (center of NW^1/$_4$ sec. 12, T. 30 N., R. 13 E.).

103. Establishment Creek Drill Hole (ECDH No. 42), 6.5 km northeast of Bloomsdale, Ste. Genevieve County, Missouri (NW^1/$_4$ NW^1/$_4$ NE^1/$_4$ sec. 32, T. 39 N., R. 8 E.). Dickeyville (House Springs) K-bentonite in Kimmswick Limestone 0.8 m above the Decorah Formation at depth of 48.5 m.

104. Establishment Creek Drill Hole (ECDH No. 43), 6.3 km northeast of Bloomsdale, Ste. Genevieve County, Missouri (SE^1/$_4$ SW^1/$_4$ SE^1/$_4$ sec. 29, T. 39 N., R. 8 E.). Dickeyville (House Springs) K-bentonite in Kimmswick Limestone 1.2 m above the Decorah Formation at depth of 51 m.

105. Laclede Gas Company No. 1 E. Mintert, St. Charles County, Missouri (1,363 ft from the south line, 2,183 ft from the east line sec. 4 T. 47 N., R. 7 E.). Deicke K-bentonite Bed at 1240 ft on gamma-ray log.

106. Drilltron, Inc. No. 2 P. Hanson, Pike County, Missouri (165 ft from north line and 175 ft from west line sec. 24, T. 52 N., R. 4 W.). Illinois State Geological Survey core C-7838.

107. Thor Resources No. 1 N. Sleight, Pike County, Illinois (NE^1/$_4$ SE^1/$_4$ NE^1/$_4$ sec. 12, T. 4 S., R. 3 W.).

108. Central Illinois Public Service Co. No. 1 Ingersoll, Pike County, Illinois (NW^1/$_4$ NW^1/$_4$ NW^1/$_4$ sec. 3, T. 4 S., R. 4 W.).

109. Central Illinois Public Service Co. No. S-1 J. G. Thomas, Brown County, Illinois (SW^1/$_4$ SE^1/$_4$ NW^1/$_4$ sec. 15, T. 1 S., R. 2 W.).

110. Nyvatex Oil Co. No. 16-1R Kearby, Schuyler County, Illinois (SW^1/$_4$ SE^1/$_4$ SE^1/$_4$ sec. 16, T. 2 N., R. 4 W.).

111. Marathon Oil Co. No. 1 Lyon Heirs, McDonough County, Illinois (1,430 ft from south line and 1,342 ft from east line NW^1/$_4$ SE^1/$_4$ sec. 19, T. 4 N., R. 4 W.).

112. New Jersey Zinc Co. H-1 R. Cochran, Hancock County, Illinois (NE^1/$_4$ SW^1/$_4$ SW^1/$_4$ SW^1/$_4$ sec. 15, T. 4 N., R. 8 W.).

113. Tylex Inc. No. 1 Marion Farms, Clark County, Missouri (4,780 ft from south line and 1,530 ft west line sec. 10, T. 65 N., R. 7 W.).

114. Natural Gas Storage Company of Illinois No. 1 H. Harris, Louisa County, Iowa (center sec. 1, T. 73 N., R. 4 W.).

115. Natural Gas Pipeline Company of America No. 1 E. U. Green, Washington County, Iowa (NE^1/$_4$ NE^1/$_4$ NE^1/$_4$ NW^1/$_4$ sec. 21, T. 76 N., R. 9 W.).

116. Big Springs BS-2, 4 km northwest of St. Olaf, Clayton County, Iowa (SE^1/$_4$ NE^1/$_4$ SE^1/$_4$ NW^1/$_4$ sec. 16, T. 94 N., R. 5 W.). Deicke at depth 76.8 m, Millbrig 75.6 m, Elkport 73.6 m.

117. Minnesota Gas Company No. 1 J. Badje, Winnebago County, Iowa (NW^1/$_4$ NW^1/$_4$ NE^1/$_4$ sec. 25, T. 100 N., R. 25 W.)

118. Northern Natural Gas Co., No. 67 1A Hollandale, Freeborn County, Minnesota (SE^1/$_4$ SE^1/$_4$ SW^1/$_4$ sec. 7, T. 103 N., R. 19 W.).

119. Drill hole—Superior Oil Company No. C-17 H. C. Ford et al., White County, Illinois (660 ft from south line and 660 ft from east line of NW^1/$_4$ SW^1/$_4$ SE^1/$_4$ sec. 27, T. 4 S., R. 14 W.). Deicke K-bentonite encountered at depth of 6560 ft 2 inches. Only 2.5 cm of K-bentonite recovered.

120. Texaco Inc. No. 1 Adam Radake, Perry County, Illinois (990 ft from north line, 840 ft from SE^1/$_4$ SW^1/$_4$ sec. 24, T. 4 S., R. 1 W.).

121. Wood Energy Inc. No. 1 Paul Borowiak, Washington County, Illinois (330 ft from south line, 330 ft from west line NE corner SE^1/$_4$ SE^1/$_4$ sec. 24, T. 3 S., R. 1 W.).

122. Texaco Inc. No. 1 R. S. Johnson, Marion County, Illinois (940 ft from south line, 335 ft from west NE^1/$_4$ NW^1/$_4$ SE^1/$_4$ sec. 6, T. 1 N., R. 2 E.).

123. C. E. Brehm No. 1 Hemminghaus Comm., Clinton County, Illinois (330 ft from north line, 380 ft from east line SW corner NE^1/$_4$ SE^1/$_4$ sec. 33, T. 3 N., R. 1 W.).

124. Illinois Power Co. No. 2 Morrell, Montgomery County, Illinois (85 ft from south line, 330 ft from west line NE corner sec. 5, T. 9 N., R. 3 W.).

125. Illinois Power Co. No. 1 Frank DeBolt, Douglas County, Illinois (846 ft from south line, 2,035 ft from west line NE corner SE^1/$_4$ sec. 4, T. 16 N., R. 8 E.).

126. Peoples Gas Light & Coke Co. No. 1 J. Lamb, De Witt County, Illinois (119 ft from south line, 1,411 ft from east line NW corner sec. 1, T. 20 N., R. 4 S.).

127. Union Hill Gas Storage Co. No. 1 J. Buchan, Piatt County, Illinois (38.5 ft from south line, 74 ft from west line of NE corner sec. 13, T. 21 N., R. 6 E.).

128. Union Hill Gas Storage Co. No. 1 Kroner, Champaign County, Illinois (32.5 ft from south line, 52.7 ft from west line NE corner sec. 32, T. 21 N., R. 7 E.).

129. Northern Illinois Gas Co. No. 1 Condit, Iroquois County, Illinois

(2,163.1 ft from south line, 175.4 ft from west line NE corner sec. 24, T. 27 N., R. 14 W.).

130. Northern Indiana Public Service Co. No. 1 H. & L. Boezman, Jasper County, Indiana (NW¼ NW¼ NW¼ sec. 6, T. 31 N., R. 7 W.).

131. U.S. Steel Co. No. 1 Gary Sheet and Tin Plant, Lake County, Indiana (SW¼ SE¼ SW¼ sec. 29, T. 37 N., R. 8 W.).

132. Security Oil & Gas Co. No. 1 Thalmann, Berrien County, Michigan (SE¼ SE¼ SE¼ sec. 10, T. 6 S., R. 17 W.). Wireline log used in cross section.

133. C. A. Perry & Son No. 1 Wooden, Cass County, Michigan (SE¼ NE¼ NW¼ sec. 8, T. 7 S., R. 14 W.). Wireline log used on cross section.

134. Upjohn Company No. 3 Upjohn, Kalamazoo County, Michigan (sec. 14, T. 3 S., R. 11 W.). Deicke, Millbrig, and Sherman Falls K-bentonite.

135. Total Petroleum Inc. No. 2-12 Emil Faist, Jackson County, Michigan (NW¼ SW¼ NE¼ sec. 12, T. 1 S., R. 1 W.). Core sample of K-bentonite from the Trenton–Black River contact, at level of 5,231 ft. Sample consists of 18 cm of mixed pale green shale and biotite-rich K-bentonite overlying 3 cm of cherty limestone. Probable Millbrig K-bentonite.

136. Mobil Oil Corp. No. 1 Gladys Kelly, Eaton County, Michigan (W½ NE¼ NW¼ sec. 24, T. 2 N., R. 3 W.).

137. McClure Oil Co. No. 1-8 Sparks, Eckelbarger, and Whightsil, Gratiot County, Michigan (NE¼ SW¼ NW¼ sec. 8, T. 10 N., R. 2 W.).

138. Humble Oil & Refining Co. No. 1 Hoppinthal, Sanilac County, Michigan (C NE¼ NE¼ sec. 16, T. 9 N., R. 15 E.).

139. Panhandle Eastern Pipeline Company No. 1 Ford Motor Company, Wayne County, Michigan (sec. 19, T. 2 S., R 11 E.).

140. Imperial Oil Limited No. 769 Imperial Calvan Malden, near Windsor, Essex County, Ontario (Malden Township, Lot 75, Concession VI, 725 ft north of south road, 1,430 ft west of east road, latitude 42° 05′ 58″, longitude 83° 01′ 27″).

141. Consumers Gas Co. No. 33556, Tract 8, Concession 4SMR, Orford Township, Kent County, Ontario.

142. Consumers Gas Co. No. 33662, Tract 1, Concession 9 I, Yarmouth Township, Elgin County, Ontario.

143. Cities Service et al. Corunna Petroleum Ltd., Lot 27, Concession VII, Dereham Township, Oxford County, Ontario, latitude 42° 54′ 43″, longitude 80° 53′ 53″).

144. Portree Shelter, Tract 8, Lot 20, Concession IV, Blenheim Township, Oxford County, Ontario.

145. Amoco A-1, Tract 2, lot 31, Concession V, Kincardine Township, Bruce County, Ontario.

146. Stanley Energy No. 1 Rickert, Monroe County, Illinois (NE¼ W¼ SE¼ sec. 32, T. 3 S., R. 9 W.).

147. Mississippi River Transmission Co. No. 1 Meisner, Perry County, Missouri (sec. 2, T. 34 N., R. 13 E.).

148. Shell Oil Co. No. 1 Trail of the Tears State Park, Cape Girardeau, Missouri (SE¼ SW¼ NW¼ NE¼ sec. 15, T. 32 N., R. 14 E.).

149. Texas Pacific Oil Co. No. 1 B. Farley, Johnson County, Illinois (680 ft from north line, 730 ft from west line NW corner SE¼ sec. 34, T. 13 S., R. 3 E.).

150. Texaco Inc. No. 1 J. M. Walters, Gallatin County, Illinois (330 ft from north line, 330 ft from east line SW¼ NE¼ SW¼ sec. 29, T. 9 S., R. 9 E.). Encountered the Deicke K-bentonite at depth of 6,476 ft.

151. General Electric Co. No. 2 General Electric Waste Disposal, Posey County, Indiana (1,113.76 ft from north line, 1.08 ft from west line NW¼ NW¼ sec. 19, T. 7 S., R. 13 W.).

152. Eastern Natural Gas Corp. No. 1 Peabody Coal Co., Warrick County, Indiana (632 ft from south line, 1,755 ft from east line SE¼ sec. 20, T. 4 S., R. 8 W.).

153. Citizens Resources Development Co. No. 1 E. Galey, Perry County, Indiana (660 ft from south line, 660 ft from west line SW¼ SE¼ sec. 5, T. 4 S., R. 1 N.).

154. Louisville Gas & Electric Co. No. 1 Joe Clark, Oldham County, Kentucky (1,080 ft from south line and 420 ft from west line, sec.12-W-50).

155. Ashland Exploration Inc. No. 1 G. Sullivan, Switzerland County, Indiana (330 ft from south line, 330 ft from east line SE¼ SW¼ sec. 20, T. 2 N., R. 2 W.).

156. Commonwealth Gas Corp. No. 1 George and Ella Covert, Jefferson Township, Adams County, Ohio.

157. Commonwealth Gas Corp. No. 1 D. P. Newell, Greenup County, Kentucky (3,850 ft from north line, 420 ft from east line sec.7-Z-78).

158. Columbia Gas Transmission Corp. No. 9784 J. H. Evans Farm, Johnson County, Kentucky (2,200 ft from south line, 880 ft from east line sec. 10-R-79).

159. Kentucky West Virginia Gas No. 7314 W. S. Dallarhide (TR#51521), Johnson County, Kentucky (775 ft from the south line and 500 ft from the east line sec. 11-P-81).

160. Signal Oil and Gas Co. No. 1 Hall, Floyd County, Kentucky (2,490 ft from north line and 2,360 ft from east line sec. 1-L-81).

161. Signal Oil & Gas Co. No. 1 Henry Stratton, Pike County, Kentucky (480 ft from north line, 1,040 ft from west line sec. 8-L-85).

162. Columbia Gas Transmission Corp. No. 20321-T, Mullins, Dickenson County, Virginia (37° 16′ 43″ N; 82° 15′ 32″ W).

163. Gulf Oil Corp. No. 1 W. R. Price, Russell County, Virginia (2,700 ft from south line, 500 ft from east line sec. 15-E-88).

164. Shell Oil Co. No. 1, L. S. Bales, Lee County, Virginia (36° 73′ 01″ N; 83° 21′ 14″ W; Virginia coordinates 140,075 N. and 575,900 E.).

165. United Fuel Gas Co. No. 9060-T Alice Shepherd, Lewis County, Kentucky, sec. 19-W-75.

166. Ashland Exploration, Inc. No. 1 Millard F. Cable, Lee County,

Kentucky, (2,310 ft from south line and 1,690 ft from west line, sec. 16-O-70).

167. Monitor Petroleum Corp. No. 1 Stanley Neeley, Jackson County, Kentucky (1,300 ft from south line and 700 ft from west line, sec. 12-L-67).

168. Triway Enterprises, Inc. No. 2A Gilbert Hensley, Clay County, Kentucky (2,440 ft from north line and 1,875 ft from west line, sec. 11-J-69).

169. Petroleum Exploration Co. No. 2 Abe Carnes, Knox County, Kentucky (2,770 ft from south line and 1,250 ft from west line, sec. 3-D-69).

170. United Fuel Gas Co. No. 8801T James Knuckles, Bell County, Kentucky (4,500 ft from north line and 1,250 ft from west line, sec. 5-C-71).

171. Comanche Oil Corp. No. 1 Petry Comm., Jackson County, Illinois (660 ft from north line, 660 ft from east line SW corner NE^1/$_4$ sec. 11, T. 7 S., R. 2 W.).

172. Victor R. Gallagher No. 1 Old Ben Coal, Williamson County, Illinois (470 ft from north line, 500 ft from east line SW^1/$_4$ SW^1/$_4$ NW^1/$_4$ sec. 1, T. 8 S., R. 1 E.).

173. Texaco Inc. No. 1 E. Cuppy, Hamilton County, Illinois (330 ft from south line, 738 ft from west line NE corner SE^1/$_4$ SW^1/$_4$ sec. 6, T. 6 S., R. 7 E.).

174. Union Oil Company of California No. 1 Cisne Community, Wayne County, Illinois (330 ft from north line, 330 ft from east line SW corner NE^1/$_4$ NE^1/$_4$ sec. 3, T. 1 S., R. 7 E.).

175. Atlantic Richfield Co. No. 77 J. B. Lewis, Lawrence County, Illinois (1,600 ft from north line, 1,422 ft from east line SW^1/$_4$ sec. 29, T. 4 N., R. 12 N.).

176. Hope Oil Limited Partnership No. 1 Patricia Harlan, Vigo County, Indiana (330 ft from south line, 330 ft from west line SW^1/$_4$ SW^1/$_4$ SW^1/$_4$ sec. 30, T. 11 N., R. 9 W.).

177. Buttercup Energy Ltd. No. 1 Rachel Pensinger, Clay County, Indiana (819 ft from north line, 110 ft from east line NE^1/$_4$ SE^1/$_4$ sec. 32, T 10 N., R. 6 W.).

178. Citizens Gas and Coke Utility No. 9 Lucille Rollison, Greene County, Indiana (690 ft from north line, 990 ft from west line NW^1/$_4$ NE^1/$_4$ sec. 8, T. 6 N., R. 5 W.).

179. Indiana Farm Bureau Co-Op Association No. 1 Luther Brown, Lawrence County, Indiana (330 ft from south line and 660 ft from east line SE^1/$_4$ SE^1/$_4$ sec. 20, T. 5 N., R. 2 E.).

180. D. B. Baxter No. 1 Darrel Richardson, Decatur County, Indiana (397 ft from south line, 330 ft from west line of SW^1/$_4$ sec. 34 T. 9 N., R. 8 E.).

181. Armco Steel Corp. No. 1 Armco Steel Corp., Butler County, Ohio (1,054.6 ft from north line, 64.7 ft from west line NW^1/$_4$ sec. 8, Lemon Township).

182. Stocker & Sitler Inc. No. 1 Coy, Clinton County, Ohio (5,161 ft from west line, 1,280 ft from east line of VMS 743, Wilson Township).

183. Ohio Department of Natural Resources No. ODS-2626, Highland County, Ohio (2,600 ft from south line, 4,300 ft from west line, Concord Township).

184. Crest Oil Co. No. 1 Clark, Ross County, Ohio (3,200 ft from south line, 8,200 ft from west line, Concord Township).

185. Clark Oil & Refining Co. No. 1 J. Huffines, Fairfield County, Ohio (660 ft from north line, 230 ft from west line NE^1/$_4$ sec. 33, Clear Creek Township).

186. Kentucky Drilling No. 4 Bartholomew, Hocking County, Ohio (955 ft from south line, 675 ft from west line Falls Township).

187. Columbia Gas No. 11043 Campbell, Morgan County, Ohio (1,300 ft from south line, 650 ft from east line, Deerfield Township).

188. Carter-Jones Lumber No. 1 Archibald, Noble County, Ohio (1,360 ft from north line, 750 ft from west line, Olive Township).

189. Ohio Department of Natural Resources No. 2621, Highland County, Ohio (6,100 ft from west line, 300 ft from north line Dodson Township).

190. Ohio Department of Natural Resources No. 3019 Cincinnati, Hamilton County, Ohio (state coordinates X: 1,418,900; Y: 409,050).

191. Ashland Oil & Refining Co. No. 1 Harold Wilson et al., Campbell County, Kentucky (1,330 ft from south line and 1,910 ft from west line 25-DD-62).

192. Pennzoil Co. No. 1 Fannie Mays, Rowan County, Kentucky (1,750 ft from south line, 1,590 ft from west line sec. 25-U-75).

193. Quasar Inc. No. 1 Sid Vice, Bath County, Kentucky (1,100 ft from south line, 650 ft from east line sec. 20-V-68).

194. Foulkner Drilling Co. No. 1 John Greenwade, Montgomery County, Kentucky (1,525 ft from north line, 1,750 ft from east line sec. 20-S-68).

195. Spanish Energy Development Corp. No. 1 Reed Unit, Lincoln County, Kentucky (2,390 from north line, 295 ft from west line sec. 21-L-58).

196. Ohio-Kentucky Oil Corp. No. 1 James Woolridge, Russell County, Kentucky (2,520 ft from north line, 2,050 ft from east line sec. 4-E-53).

197. W. K. Griffin Jr. No. 1 Anna Mary Creekmore, McCreary County, Kentucky (1,200 ft from south line, 75 ft from east line sec. 8-B-62).

198. Leland Petroleum Production Co. No. 1 R. C. Hutchins, Morgan County, Tennessee (sec. 24-3S-57E).

199. B & W Oil Co. No. 1 Geer, White County, Tennessee (1,350 ft from south line, 100 ft from east line sec. 15-6S-49E).

200. Hickory Creek Development Corp. No. 1 John K. Tart, Coffee County, Tennessee (700 ft from north line, 1,120 ft from east line sec. 25-12S-43E).

201. United American Energy No. 1 Modena, Franklin County, Tennessee (1,800 ft from north line, 1,520 ft from east line sec. 16-16S-42E).

202. Southland Royalty Co. No. 1 Allison, Jackson County, Alabama (660 ft from north line, 660 ft from east line sec. 27-2S-7E).

203. Cincinnati Gas and Electric No. 1 Thomason, Gallatin County, Kentucky (2,820 ft from the north line, 840 ft from the east line, sec. 8-AA-57). Unnamed K-bentonites in Knox Group at drilling depth of 788 ft and 831 ft.

204. Cities Service Minerals Corp. No. B.T.3 O'Donovan, Owen County, Kentucky (2,780 ft from the north line, 110 ft from the west line, sec. 7-Y-55). Unnamed K-bentonites in Knox Group at drilling depth of 778 ft and 826 ft.

205. Jarvis and Marcell No. 1 Parrigan, Clinton County, Kentucky. K-bentonite described by Freeman (1953) from the Wells Creek Dolomite at depth of 1,752 ft to 1,757 ft.

206. Earth Science Laboratories, Inc. No. 2 Thomason, Gallatin County, Kentucky, sec. 8-AA-57. Unnamed K-bentonites in Knox Group at drilling depth of 788 ft and 832 ft.

207. Ontario Geological Survey No. 82-2 Harwich 25-I ECR drill hole, Lot 25, Tract 1, Concession I ECR, Harwich Township, Chatham, Kent County, Ontario. K-bentonites at 1025.5 m to 1025.7 m and 1031.1 m to 1031.4 m tentatively identified by Trevail (1990) as the Millbrig and Deicke, respectively.

208. Manitoulin Island DDH No. 2 Elizabeth Bay, Manitoulin County (Bidwell Township, precise location unavailable from the Ontario Geological Survey).

209. ARCO No. 1-14 Dunn, Calhoun County, Michigan (SE$^{1}/_{4}$ SE$^{1}/_{4}$ SE$^{1}/_{4}$ sec. 14, T. 3 S., R. 5 W.). Core sample (MI-1) through the "Black River shale" at depth of 4,144 ft.

210. Sovereign Exploration Co. No. 1 Post, Seminole County, Oklahoma (C NE$^{1}/_{4}$ NE$^{1}/_{4}$ NE$^{1}/_{4}$ sec. 8, T. 8 N., R. 6 E.). Probable K-bentonites at top and base of Corbin Ranch "submember," Pooleville Limestone Member, Bromide Formation.

211. Cicada Petroleum Corp. No. 1 Dad, Pottawatomie County, Oklahoma (NE$^{1}/_{4}$ NW$^{1}/_{4}$ NE$^{1}/_{4}$ sec. 19, T. 7 N., R. 5 E.). Probable K-bentonites at top and base of Corbin Ranch "submember," Pooleville Limestone Member, Bromide Formation.

212. Empire Oil Co. No. 1 Schwalm, Wabaunsee County, Kansas (SW$^{1}/_{4}$, sec. 19, T. 12 S., R. 11 E.). Drill cuttings described by Taylor (1947). Possible K-bentonite beds in strata equivalent to the Platteville and Decorah of the Mississippi Valley.

213. Iowa Geological Survey No. 1 Camp Quest (drill core), 3 km northeast of Le Mars, Plymouth County, Iowa (SW$^{1}/_{4}$ sec. 2, T. 92 N., R. 45 W.). Deicke K-bentonite in upper part of the Platteville Group identified by chemical fingerprinting (Kolata et al., 1986).

214. Quarry on west side of county road, 6 km south of Annaton, Grant County, Wisconsin (SE$^{1}/_{4}$ SW$^{1}/_{4}$ SE$^{1}/_{4}$ sec. 25, T. 5 N., R. 2 W.). Exposure of the Deicke and Millbrig K-bentonite Beds. Annaton South Section (Kolata et al., 1986).

215. Roadcut on north side of U.S. Highway 18, 5 km east of intersection of highways 18 and 23, Dodgeville, Iowa County, Wisconsin (NE$^{1}/_{4}$ NW$^{1}/_{4}$ SW$^{1}/_{4}$ sec. 19, T. 6 N., R. 4 E.). Exposure of the Dickeyville K-bentonite Bed. Dodgeville East Section (Kolata et al., 1986).

216. Kapper Construction Company "Grabau Quarry" situated 6.4 km east and 5.6 km south of Spring Valley, Fillmore County, Minnesota (SE$^{1}/_{4}$ SE$^{1}/_{4}$ sec. 17, T. 102 N., R. 12 W.). Wise Lake and Dubuque Formations of the Galena Group. Two unnamed K-bentonite beds in the Dubuque Formation. Unpublished section M-100 (Calvin O. Levorson and Arthur J. Gerk, 1973).

217. Roadcut on west side of Bald Hill Road (State Route W) 2 km south of Eureka, St. Louis County, Missouri (SW$^{1}/_{4}$ SW$^{1}/_{4}$ SE$^{1}/_{4}$ irregular sec. 1, T. 43 N., R. 3 E.). Exposure of the Deicke, Millbrig, and Dickeyville K-bentonite Beds. Eureka South Section (Kolata et al., 1986).

218. Roadcut on east side of State Highway 30, 0.8 km north of intersection with Highway W at House Springs, Jefferson County, Missouri (irregular unnumbered section between sec. 32 and sec. 33, T. 43 N., R. 4 E.). Exposure of the Deicke, Millbrig, and Dickeyville K-bentonite Beds. House Springs Section (Kolata et al., 1986).

219. Missouri Limestone Company Quarry, 6.4 km southeast of Warrenton, Warren County, Missouri (SE$^{1}/_{4}$ NE$^{1}/_{4}$ NE$^{1}/_{4}$ sec. 15, T. 46 N., R. 2 W.). Exposure of the Deicke and Millbrig K-bentonite Beds. Warrenton South Section (Kolata et al., 1986).

220. Small abandoned quarry on dead-end road, 400 m northeast of intersection of U.S. Highways 127 and 421, on northwest side of Frankfort, Franklin County, Kentucky. Exposure of the Deicke K-bentonite Bed.

221. Pennsy Supply Company Quarry situated immediately north of Route 11 (Carlisle Pike), 12.8 km west of Harrisburg, Cumberland County, Pennsylvania (Mechanicsburg 7.5' Quadrangle). Exposure of late Mohawkian K-bentonites in the Oranda Formation.

222. Natural outcrop at Trenton Falls on West Canada Creek south of Prospect, Oneida County, New York. UTM coordinates provided courtesy of Charles E. Mitchell as follows: bottom of stratigraphic section at 4,790,732 N and 487,049 E; top of section at 4,791,902 N and 487,488 E.

REFERENCES CITED

Adams, J. A. S., Osmond, J. K., Edward, G., and Henle, W., 1960, Absolute dating of the Middle Ordovician: Nature, v. 188, p. 636–638.

Ahn, J. H., and Buseck, P. R., 1990, Layer-stacking sequences and structural disorder in mixed-layer illite/smectite: Image simulations and HRTEM imaging: American Mineralogist, v. 75, p. 267–275.

Alberstadt, L., and Repetski, J. E., 1989, A Lower Ordovician sponge/algal facies in the southern United States and its counterparts elsewhere in North America: Palaios, v. 4, p. 225–242.

Allen, V. T., 1929, Altered tuffs in the Ordovician of Minnesota: Journal of Geology, v. 37, no. 3, p. 239–248.

Allen, A. T., and Lester, J. G., 1957, Zonation of the Middle and Upper Ordovician strata in northwestern Georgia: Geological Survey of Georgia, Bulletin 66, 110 p.

Altaner, S. P., 1989, Calculation of K diffusional rates in bentonite beds: Geochimica et Cosmochimica Acta, v. 53, p. 923–931.

Altaner, S. P., and Bethke, C. M., 1989, Interlayer order in illite/smectite: American Mineralogist, v. 73, p. 766–774.

Amsden, T. W., and Sweet, W. C., 1983, Upper Bromide Formation and Viola Group (Middle and Upper Ordovician) in eastern Oklahoma: Oklahoma Geological Survey Bulletin, no. 132, 76 p.

Aronson, J. L., and Lee, M., 1986, K/Ar systematics of bentonite and shale in a contact metamorphic zone, Cerrillos, New Mexico: Clays and Clay Minerals, v. 4, p. 483–487.

Atherton, E., 1971, Tectonic development of the Eastern Interior region of the United States: Illinois State Geological Survey, Illinois Petroleum, v. 96, p. 29–43.

Barnes, C. R., Norford, B. S., Skevington, D., 1981, The Ordovician System in Canada: Ottawa, Ontario, International Union of Geological Sciences, Publication No. 8, 27 p.

Barnes, C. R., Telford, P. G., and Tarrant, G. A., 1978, Ordovician and Silurian conodont biostratigraphy, Manitoulin Island and Bruce Peninsula, Ontario, in Sanford, J. T., and Mosher, R. E., eds., Geology of the Manitoulin area: Lansing, Michigan, Michigan Basin Geological Society, Special Papers no. 3, p. 63–71.

Barnett, S. G., III, 1965, Conodonts of the Jacksonburg Limestone (Middle Ordovician) of northwestern New Jersey and eastern Pennsylvania: Micropaleontology, v. 11, no. 1, p. 59–80.

Bates, R. L., 1939, Geology of Powell Valley in northwestern Lee County, Virginia: Virginia Geological Survey Bulletin 51-B, p. 31–94.

Bell, W. C., 1954, Upper Mississippi Valley Middle Ordovician bentonites: Geological Society of America Abstracts with Program, v. 65, p. 1230–1231.

Belt, E. S., Riva, J., and Bussières, L., 1979, Revision and correlation of late Middle Ordovician stratigraphy northeast of Quebec City: Canadian Journal of Earth Sciences, v. 16, p. 1467–1483.

Bergström, S. M., 1971, Conodont biostratigraphy of the Middle and Upper Ordovician of Europe and eastern North America: Boulder, Colorado, Geological Society of America Memoir 127, p. 83–157.

Bergström, S. M., 1973, Biostratigraphy and facies relations in the lower Middle Ordovician of easternmost Tennessee: American Journal of Science, v. 273, p. 261–293.

Bergström, S. M., 1989, Use of graphic correlation for assessing event-stratigraphic significance and trans-Atlantic relationships of Ordovician K-bentonites, in Proceedings, Estonian Academy of Sciences, v. 38: Tallinn, Estonia, Estonian Academy of Sciences, p. 55–59.

Bergström, S. M., 1990, Biostratigraphic and biogeographic significance of Middle and Upper Ordovician conodonts in the Girvan succession, south-west Scotland: Forschungsinstitut Senckenberg Courier, v. 118, p. 1–43.

Bergström, S. M., Carnes, J. B., Hall, J. C., Kurapkat, W., and O'Neill, B. E., 1988, Conodont biostratigraphy of some Middle Ordovician stratotypes in the southern and central Appalachians: New York State Museum Bulletin, v. 462, p. 20–32.

Bergström, S. M., and Mitchell, C. E., 1992, The Ordovician Utica Shale in the eastern Midcontinent region: Age, lithofacies, and regional relationships, in Chaplin, J. R., and Barrick, J. E., eds., Special papers in paleontology and stratigraphy: A tribute to Thomas W. Amsden: Oklahoma Geological Survey Bulletin 145, p. 67–89.

Bergström, S. M., and Mitchell, C. E., 1994, Regional relationships between late Middle and early Late Ordovician standard successions in New York and Quebec and the Cincinnati region in Ohio, Indiana, and Kentucky, in Landing, E., Studies in stratigraphy and paleontology in honor of Donald W. Fisher: New York State Museum Bulletin 481, p. 5–20.

Bergström, S. M., Huff, W. D., Kolata, D. R., Bauert, H., 1995, Nomenclature, stratigraphy, chemical fingerprinting, and areal distribution of some Middle Ordovician K-bentonites in Baltoscandia: Geologiska Föreningens i Stockholm Förhandlingar, v. 117, p. 1–13.

Bohor, B. F., and Triplehorn, D. M., 1993, Tonsteins: Altered volcanic-ash layers in coal-bearing sequences: Boulder, Colorado, Geological Society of America Special Paper 285, 44 p.

Bonine, C. A., and Honess, A. P., 1929, Bentonite in Pennsylvania: Pennsylvania Academy of Scientific Proceedings, v. 3, p. 18–25.

Borchardt, G. A., Harward, M. E., and Schmitt, R. A., 1971, Correlation of volcanic ash deposits by activation analysis of glass separates: Quaternary Research, v. 1, p. 247–260.

Brun, J., and Chagnon, A., 1979, Rock stratigraphy and clay mineralogy of volcanic ash beds from the Black River and Trenton Groups of southern Quebec: Canadian Journal of Earth Sciences, v. 16, p. 1499–1507.

Brusewitz, A. M., 1986, Chemical and physical properties of Paleozoic potassium bentonites from Kinnekulle, Sweden: Clays and Clay Minerals, v. 34, p. 442–454.

Brusewitz, A. M., 1988, Asymmetric zonation of a thick Ordovician K-bentonite bed at Kinnekulle, Sweden: Clays and Clay Minerals, v. 36, p. 349–353.

Butts, C., 1926, The Paleozoic rocks, in Adams, G. I., ed., The geology of Alabama: Alabama Geological Survey Special Report 14, p. 41–230.

Cady, G. H., 1919, Geology and mineral resources of the Hennepin and LaSalle Quadrangles: Illinois State Geological Survey Bulletin 37, 136 p.

Calvert, W. L., 1962, Sub-Trenton rocks from Lee County, Virginia, to Fayette County, Ohio: Ohio Geological Survey Report of Investigations, no. 45, 57 p.

Calvert, W. L., 1963a, A cross section of sub-Trenton rocks from Wood County, West Virginia to Fayette County, Illinois: Ohio Geological Survey Report of Investigations, no. 48, 33 p.

Calvert, W. L., 1963b, Sub-Trenton rocks of Ohio in cross sections from West Virginia and Pennsylvania to Michigan: Ohio Geological Survey Report of Investigations, no. 49, 5 p.

Calvert, W. L., 1964, Sub-Trenton rocks from Fayette County, Ohio to Brant County, Ontario: Ohio Geological Survey Report of Investigations, no. 52, 7 p.

Carter, B. D., and Chowns, T. M., 1986, Stratigraphic and environmental relationships of Middle and Upper Ordovician rocks in northwestern Georgia and northeastern Alabama, in Benson, D. J., and Stock, C. W., eds., Depositional history of the Middle Ordovician of the Alabama Appalachians: Tuscaloosa, Alabama, The Alabama Geological Society, 23rd Annual Field Trip Guidebook, p. 33–50.

Caster, K. E., and Kjellesvig-Waering, E. N., 1964, Upper Ordovician Eurypterids of Ohio: Paleontographica Americana, v. 4, no. 32, p. 301–358.

Catacosinos, P. A., Daniels, P. A., Jr., Harrison, W. B., III., 1991, Structure, stratigraphy, and petroleum geology of the Michigan Basin, in Leighton, M. W., Kolata, D. R., Oltz, D. F., and Eidel, J. J., eds., Interior cratonic basins: Tulsa, Oklahoma, American Association of Petroleum Geologists Memoir 51, p. 561–601.

Cawood, P. A., Barnes, C. R., Botsford, J. W., James, N. P., Knight, I., O'Brien, S. J., O'Neill, P. P., Parsons, M. G., Stenzel, S. R., Stevens, R. K., and Williams, S. H., 1988, A cross-section of the Iapetus Ocean and its continental margins, in Williams, S. H., ed., Field Excursion Guidebook: St. John's, Newfoundland, 5th, International Symposium on the Ordovician System.

Cetin, K., and Huff, W. D., 1995, Layer charge and smectite-like character of the expandable component of illite/smectite in K-bentonite as determined by alkylammonium ion exchange: Clays and Clay Minerals, v. 43, no. 2, p. 150–158.

Cheng, Y., and Pettry, D. E., 1993, Horizontal and vertical movements of two expansive soils in Mississippi: Soil Science Society of America Journal, v. 57, p. 1542–1547.

Cisne, J. L., and Chandlee, G. O., 1982, Taconic foreland basin graptolites: Age zonation, depth zonation, and use in ecostratigraphic correlation: Lethaia, v. 15, p. 343–363.

Cisne, J. L., and Rabe, B. D., 1978, Coenocorrelation: Gradient analysis of fossil communities and its applications in stratigraphy: Lethaia, v. 11, p. 341–364.

Cisne, J. L., Karig, D. E., Rabe, B. D., and Hay, B. J., 1982, Topography and tectonics of the Taconic outer trench slope as revealed through gradient analysis of fossil assemblages: Lethaia, v. 15, p. 229–246.

Cloetingh, S., McQueen, H., Lambeck, K., 1985, On a tectonic mechanism for regional sea-level variations: Earth and Planetary Science Letters, v. 75, p. 157–166.

Coker, A. E., 1962, A mineralogical study of an Ordovician metabentonite near Clinton, Anderson County, Tennessee [M.S. thesis]: Knoxville, University of Tennessee, 49 p.

Conkin, J. E., 1991, Middle Ordovician (Mohawkian) paracontinuous stratigraphy and metabentonites of eastern North America: University of Louisville Studies in Paleontology and Stratigraphy no. 18, 54 p.

Conkin, J. E., and Conkin, B. M., 1973, The paracontinuity and the determination of the Devonian-Mississippian boundary in the type lower Mississippian area of North America: University of Louisville Studies in Paleontology and Stratigraphy, no. 1, 36 p.

Conkin, J. E., and Conkin, B. M., 1983, Paleozoic metabentonites of North America: Part 2—Metabentonites in the Middle Ordovician Tyrone Formation at Boonesborough, Clark County, Kentucky: University of Louisville Studies in Paleontology and Stratigraphy, no. 17, 47 p.

Conkin, J. E., and Conkin, B. M., 1992, Paleozoic metabentonites of North America: Part 3—New Ordovician metabentonites from Kentucky and Tennessee: University of Louisville Studies in Paleontology and Stratigraphy no. 20, 30 p.

Conkin, J. E., and Dasari, M. R., 1986, Capitol Metabentonite in the Trenton Curdsville Limestone of central Kentucky: University of Louisville Notes in Paleontology and Stratigraphy B, 15 p.

Cook, T. D., and Bally, A. W., 1975, Stratigraphic Atlas of North and Central America: Princeton, New Jersey, Princeton University Press, 272 p.

Cooper, B. N., and Cooper, G. A., 1946, Lower Middle Ordovician stratigraphy of the Shenandoah Valley, Virginia: Geological Society of America Bulletin, v. 57, p. 35–113.

Cousineau, P. A., 1994, Subaqueous pyroclastic deposits in an Ordovician forearc basin: An example from the Saint-Victor Formation, Quebec Appalachians, Canada: Journal of Sedimentary Research, v. A64, p. 867–880.

Craddock, J. P., Jackson, M., van der Pluijm, B. A., and Versical, R. T., 1993, Regional shortening fabrics in eastern North America: Far-field stress transmission from the Appalachian-Ouachita orogenic belt: Tectonics, v. 12, p. 256–264.

Cressman, E. R., 1973, Lithostratigraphy and depositional environments of the Lexington Limestone (Ordovician) of central Kentucky: U.S. Geological Survey Professional Paper 768, 61 p.

Cressman, E. R., and Noger, M. C., 1976, Tidal-flat carbonate environments in the High Bridge Group (Ordovician) of central Kentucky: Kentucky Geological Survey Series X, Report of Investigation 18, 15 p.

Cullen-Lollis, J., and Huff, W. D., 1986, Correlation of Champlainian (Middle Ordovician) K-bentonite beds in central Pennsylvania based on chemical fingerprinting: Journal of Geology, v. 94, p. 865–874.

Curry, B. B., Graese, A. M., Hasek, M. J., Vaiden, R. C., Bauer, R. A., Schumacher, D. A., Norton, K. A., and Dixon, W. G., Jr., 1988, Geological-geotechnical studies for siting the Superconducting Super Collider in Illinois: Results of the 1986 test drilling program: Illinois State Geological Survey, Environmental Geology Notes 122, 108 p.

Decker, C. E., 1933, Viola Limestone, primarily of Arbuckle and Wichita Mountain regions, Oklahoma: American Association of Petroleum Geologists Bulletin, v. 17, no. 12, p. 1405–1435.

Delano, J. W., 1992, Chronostratigraphy of the Trenton Group and Utica Shale, Pt. II: Stratigraphic correlations using Ordovician glasses in K-bentonites: Geological Society of America Abstracts with Programs, v. 24, no. 7, p. 197.

Delano, J. W., Schirnick, C., Bock, B., Kidd, W. S. F., Heizler, M. T., Putman, G. W., DeLong, S. E., and Ohr, M., 1990, Petrology and geochemistry of Ordovician K-bentonites in New York State: Constraints on the nature of a volcanic arc: Journal of Geology, v. 98, p. 157–170.

Delano, J. W., Tice, S. J., Mitchell, C. E., and Goldman, D., 1994, Rhyolitic glass in Ordovician K-bentonites: A new stratigraphic tool: Geology, v. 22, p. 115–118.

Dickerson, P. W., 1994, Ordovician tectonics, sedimentation and volcanism of southern Laurentian margin—southwestern United States and eastern Mexico: Joint meeting of IGCP Projects 376 and 319 and the 2nd Circum-Atlantic, Circum-Pacific conference, Nova Scotia, Program and Abstracts, p. 15.

Diecchio, R. J., 1991, Taconian sedimentary basins of the Appalachians, in Barnes, C. R., and Williams, S. H., eds., Advances in Ordovician geology: Geological Survey of Canada, Paper 90-9, p. 225–234.

Dodson, M. H., 1979, Theory of cooling ages, in Ger, E., and Hunziker, J. C., eds., Lectures in isotope geology: New York, New York, Springer-Verlag, p. 194–202.

Dodson, M. H., 1973, Closure temperature in cooling geochronological and petrological systems: Contributions to Mineralogy and Petrology, v. 40, p. 259–274.

Drahovzal, J. A., and Neathery, T. L., 1971, The Middle and Upper Ordovician of the Alabama Appalachians: Alabama Geological Society, Guidebook to the 9th Annual Field Trip, 229 p.

Drake, A. A., Jr., Sinha, A. K., Laird, J., and Guy, R. E., 1989, The Taconic orogen, in Hatcher, R. D., Jr., Thomas, W. A., and Viele, G. W., eds., The Appalachian-Ouachita Orogen in the United States: Boulder, Colorado, Geological Society of America, Geology of North America, v. F-2, p. 101–177.

Elliott, W. C., and Aronson, J. L., 1987, Alleghanian episode of K-bentonite illitization in the southern Appalachian Basin: Geology, v. 15, p. 735–739.

Elliott, W. C., and Aronson, J. L., 1993, The timing and extent of illite formation in Ordovician K-bentonites at the Cincinnati Arch, the Nashville Dome and northeastern Illinois Basin: Basin Research, v. 5, p. 125–135.

Ettensohn, F. R., 1991, Flexural interpretation of relationships between Ordovician tectonism and stratigraphic sequences, central and southern Appalachians, U.S.A., in Barnes, C. R., and Williams, S. H., eds., Advances in Ordovician geology: Geological Survey of Canada, Paper 90-9, p. 213–224.

Faber, J. A., 1979, Geomagnetic reversal stratigraphy and conodont biostratigraphy of the Middle Ordovician Trenton Limestone, Michigan Basin [M.S. thesis]: Columbus, The Ohio State University, 80 p.

Faul, H., 1960, Geologic time scale: Geological Society of America Bulletin, v. 71, p. 637–644.

Faul, H., and Thomas, H., 1959, Argon ages of the great ash bed from the Ordovician of Alabama and of the bentonite marker shale from Tennessee: Geological Society of America Bulletin, v. 70, p. 1600–1601.

Fay, R. O., and Graffham, A. A., 1982, Measured sections and collecting localities, in Sprinkle, J., ed., Echinoderm faunas from the Bromide Formation (Middle Ordovician) of Oklahoma: The University of Kansas Paleontological Contributions, Monograph 1, p. 335–369.

Fetzer, J. A., 1973, Biostratigraphic evaluation of some Middle Ordovician bentonite complexes in eastern North America [M.S. thesis]: Columbus, The Ohio State University, 160 p.

Finney, S. C., 1982, Ordovician graptolite zonation, in Ross, R. J., Jr., The Ordovician system in the United States: International Union of Geological Sciences, Publication 12, p. 14.

Finney, S. C., 1983, Biogeography of Ordovician graptolites in the southern Appalachians, in Bruton, D. L. ed., Aspects of the Ordovician System: Paleontological Contributions from the University of Oslo, no. 295, p. 161–170.

Finney, S. C., and Skevington, D., 1979, A mixed Atlantic-Pacific province Middle Ordovician graptolite fauna in western Newfoundland: Canadian Journal of Earth Sciences, v. 16, p. 1899–1902.

Fisher, D. W., 1977, Correlation of the Hadrynian, Cambrian, and Ordovician rocks in New York State: Albany, New York State Museum, Map and Chart Series 25.

Fisher, D. W., 1982, Synopsis of Ordovician correlations in New York State, in Ross, R. J., Jr., The Ordovician System in the United States: Ottawa, Ontario, International Union of Geological Sciences, Publication 12, p. 45–51.

Fox, P. P., and Grant, L. F., 1944, Ordovician bentonites in Tennessee and adjacent states: Journal of Geology, v. 52, p. 319–332.

Freeman, L. B., 1953, Regional subsurface stratigraphy of the Cambrian and Ordovician in Kentucky and Vicinity: Kentucky Geological Survey, Bulletin 12, Series IX, 352 p.

Ghosh, P. K., 1972, Use of bentonites and glauconites in potassium-40/argon-40 dating in Gulf Coast stratigraphy [Ph.D. thesis]: Houston,

Texas, Rice University, 136 p.

Giles, A. W., 1927, The origin and occurrences in Rockbridge County, Virginia, of so-called "bentonite": Journal of Geology, v. 35, p. 527–541.

Goldman, D., and Mitchell, C. E., 1994, Three-dimensional graptolites from the upper Middle Ordovician Neuville Formation, Quebec, in Landing, E., ed., Studies in stratigraphy and paleontology in honor of Donald W. Fisher: New York State Museum Bulletin 481, p. 87–99.

Goldman, D., Mitchell, C. E., Bergström, S. M., Delano, J. W., and Tice, S., 1994, K-bentonites and graptolite biostratigraphy in the Middle Ordovician of New York State and Quebec: A new chronostratigraphic model: Palaios, v. 9, p. 124–143.

Hall, J. C., 1986, Conodont biostratigraphy and facies relationships of the Middle Ordovician in the western overthrust region and Sequatchie Valley of the southern Appalachians [Ph.D. thesis]: Columbus, The Ohio State University, 345 p.

Harris, L. D., 1967, Geology of the L. S. Bales well, Lee County, Virginia—A Cambrian and Ordovician test: Lexington, Kentucky Geological Survey, Series X, Special Publication 14, p. 50–55.

Hatch, J. R., Jacobson, S. R., Witzke, B. J., Risatti, J. B., Anders, D. E., Watney, W. L., Newell, K. D., and Vuletich, A. K., 1987, Possible late Middle Ordovician organic carbon isotope excursion: Evidence from Ordovician oils and hydrocarbon source rocks, Midcontinent and east-Central United States: American Association of Petroleum Geologists Bulletin, v. 71, p. 1342–1354.

Hatcher, R. D., Jr., 1987, Tectonics of the southern and central Appalachian internides: Annual Review of Earth and Planetary Sciences, v. 15, p. 337–362.

Hatcher, R. D., Jr., 1989, Tectonic synthesis of the U.S. Appalachians, in Hatcher, R. D., Jr., Thomas, W. A., and Viele, G. W., eds., The Appalachian-Ouachita Orogen in the United States: Boulder, Colorado, Geological Society of America, Geology of North America, v. F-2, p. 511–535.

Hay, R. L., Lee, M., Kolata, D. R., Matthews, J. C., and Morton, J. P., 1988, Episodic potassic diagenesis of Ordovician tuffs in the Mississippi valley area: Geology, v. 16, p. 743–747.

Haynes, J. T., 1992, Reinterpretation of Rocklandian (Upper Ordovician) K-bentonites stratigraphy in southwest Virginia, southeast West Virginia, and northeast Tennessee: Virginia Division of Mineral Resources, Publication 126, 58 p.

Haynes, J. T., 1994, The Ordovician Deicke and Millbrig K-bentonite Beds of the Cincinnati Arch and the Southern Valley and Ridge Province: Boulder, Colorado, Geological Society of America, Special Paper 290, 80 p.

Haynes, J. T., and Melson, W. G., 1995, Biotite composition indicates that the two Ordovician K-bentonites at the Old North Ragland Quarry, Alabama, are the same structurally repeated tephra layer: Geological Society of America Abstracts with Programs, v. 27, no. 2, p. 60–61.

Hergenroder, J. D., 1966, The Bays Formation (Middle Ordovician) and related rocks of the southern Appalachians [Ph.D. thesis]: Blacksburg, Virginia Polytechnic Institute and State University, 325 p.

Hergenroder, J. D., 1973, Stratigraphy of the Middle Ordovician bentonites in southern Appalachians: Geological Society of America Abstracts with Programs, v. 5, p. 403.

Heyman, L., 1970, Petrology of the basal Middle Ordovician Blackford Formation of type belt, Russell County, Virginia [Ph.D. thesis]: Blacksburg, Virginia Polytechnic Institute and State University, 272 p.

Hildreth, W., Halliday, A. N., and Christiansen, R. L., 1991, Isotopic and chemical evidence concerning the genesis and contamination of basaltic and rhyolitic magma beneath the Yellow Plateau volcanic field: Journal of Petrology, v. 32, p. 63–138.

Hiscott, R. N., 1978, Provenance of Ordovician deep-water sandstones, Tourelle Formation, Quebec, and implications for initiation of the Taconic Orogeny: Canadian Journal of Earth Sciences, v. 15, p. 1579–1597.

Hower, J., Eslinger, E. V., Hower, M. E., and Perry, E. A., 1976, Mechanism of burial metamorphism of argillaceous sediment: 1. Mineralogical and chemical evidence: Geological Society of America Bulletin, v. 87, p. 725–737.

Huff, W. D., 1983, Correlation of Middle Ordovician K-bentonites based on chemical fingerprinting: Journal of Geology, v. 91, p. 657–669.

Huff, W. D., and Kolata, D. R., 1989, Correlation of K-bentonite beds by chemical fingerprinting using multivariate statistics, in Cross, T. A., ed., Quantitative dynamic stratigraphy: Englewood Cliffs, New Jersey, Prentice-Hall, p. 567–577.

Huff, W. D., and Kolata, D. R., 1990, Correlation of the Ordovician Deicke and Millbrig K-bentonites between the Mississippi Valley and the southern Appalachians: American Association of Petroleum Geologists, v. 74, p. 1736–1747.

Huff, W. D., and Türkmenoglu, A. G., 1981, Chemical characteristics and origin of Ordovician K-bentonites along the Cincinnati Arch: Clays and Clay Minerals, v. 29, p. 113–123.

Huff, W. D., Whiteman, J. A., and Curtis, C. D., 1988, Investigation of a K-bentonite by X-ray powder diffraction and analytical transmission electron microscopy: Clays and Clay Minerals, v. 36, p. 83–93.

Huff, W. D., Anderson, T. B., Rundle, C. C., and Odin, G. S., 1991, Chemostratigraphy, K-Ar ages and illitization of Silurian K-bentonites from the central belt of the Southern Uplands-Down-Longford terrane, British Isles: Journal of the Geological Society, London, United Kingdom, v. 148, p. 861–868.

Huff, W. D., Bergström, S. M., and Kolata, D. R., 1992, Gigantic Ordovician volcanic ash fall in North America and Europe: Biological, tectonomagmatic, and event-stratigraphic significance: Geology, v. 20, p. 875–878.

Huff, W. D., Merriman, R. J., Morgan, D. J., and Roberts, B., 1993, Distribution and tectonic setting of Ordovician K-bentonites in the United Kingdom: Geological Magazine, v. 130, p. 93–100.

Huffman, G. G., 1945, Middle Ordovician limestones from Lee County, Virginia to central Kentucky: Journal of Geology, v. 53, p. 145–174.

Hunziker, J. C., 1979, Potassium argon dating, in Ger, E., and Hunziker, J. C., eds., Lectures in isotope geology: New York, New York, Springer-Verlag, p. 52–76.

Hunziker, J. C., 1986, The evolution of illite to muscovite: An example of the behavior of isotopes in low-grade metamorphic terrains: Chemical Geology, v. 57, p. 31–40.

Hussey, R. C., 1952, The Middle and Upper Ordovician rocks of Michigan: Michigan Geological Survey Division, Publication 46, Geological Series 39, 89 p.

Iijima, S., and Busek, P. R., 1978, Experimental study of disordered mica structures by high-resolution electron microscopy: Acta Crystallographica, v. A34, p. 709–719.

Jaanusson, C. M., and Bergström, S. M., 1980, Middle Ordovician faunal spatial differentiation in Baltoscandia and the Appalachians: Alcheringa, v. 4, p. 89–110.

Jacobi, R. D., 1981, Peripheral bulge—A causal mechanism for the Lower/Middle Ordovician unconformity along the western margin of the northern Appalachians: Earth and Planetary Science Letters, v. 56, p. 245–251.

Jagodzinski, H., 1949, Ein dimensionale Fehlordnung in Kristallen und ihr Einfluss auf die Röntgeninterferenzen. I. Berechnung des Fehlordnungsgrades aus der Röntgenintensitäten: Acta Crystallographica, v. 2, p. 201–207.

James, N. R., Barnes, C. R., Boyce, W. D., Cawood, P. A., Knight, I., Stenzel, S. R., Stevens, R. K., and Williams, S. H., 1988, Carbonates and faunas of western Newfoundland: St. John's, Newfoundland, Canada, Subcommission on Ordovician Stratigraphy (ICS/IUGS), International Commission on Stratigraphy, IGCP Project 216, 5th International Symposium on the Ordovician System, August 1988.

Kay, G. M., 1931, Stratigraphy of the Hounsfield Metabentonite: Journal of Geology, v. 39, p. 361–376.

Kay, G. M., 1934, Distribution of Ordovician altered volcanic materials and related clays, in Proceedings, Geological Society of America, 1933: New York, New York, Geological Society of America, p. 380.

Kay, G. M., 1935, Distribution of Ordovician altered volcanic materials and related clays: Geological Society of America Bulletin, v. 46, p. 225–244.

Kay, G. M., 1937, Stratigraphy of the Trenton Group: Geological Society of America Bulletin, v. 48, p. 233–302.

Kay, G. M., 1944, Middle Ordovician of central Pennsylvania: Journal of Geology, v. 52, p. 1–23, 97–116.

Kay, G. M., 1953, Geology of the Utica Quadrangle, New York, New York State Museum Bulletin, v. 347, 126 p.

Kay, G. M., 1956, Ordovician limestones in the western anticlines of the Appalachians in West Virginia and Virginia northeast of the New River: Geological Society of America Bulletin, v. 67, p. 55–106.

Keith, B. D., 1985, Facies, diagenesis, and the upper contact of the Trenton Limestone of northern Indiana, in Cercone, J. M., and Budai, J. M., eds., Ordovician and Silurian rocks of the Michigan Basin and its margin: Michigan Basin Geological Society Special Paper 4, p. 15–32.

Keith, B. D., 1989, Regional facies of the Upper Ordovician Series of eastern North America, in Keith, B. D., ed., The Trenton Group (Upper Ordovician Series) of eastern North America: Deposition, diagenesis, and petroleum: American Association of Petroleum Geologists Studies in Geology, v. 29, p. 1–16.

Kempton, J. P., Bauer, R. A., Curry, B. B., Dixon, W. G., Graese, A. M., Reed, P. C., Sargent, M. L., and Vaiden, R. C., 1987a, Geological-geotechnical studies for siting the Superconducting Super Collider in Illinois: Results of the fall 1984 test drilling program: Illinois State Geological Survey, Environmental Geology Notes 117, 102 p.

Kempton, J. P., Bauer, R. A., Curry, B. B., Dixon, W. G., Jr., Graese, A. M., Reed, P. C., and Vaiden, R. C., 1987b, Geological-geotechnical studies for siting the Superconducting Super Collider in Illinois: Results of the spring 1985 test drilling program: Illinois State Geological Survey, Environmental Geology Notes 120, 88 p.

King, P. B., 1937, Geology of the Marathon region, Texas: U.S. Geological Survey Professional Paper 187, 148 p.

Klecka, W. R., 1981, Discriminant analysis, in Nie, N. H., Hull, C. H., Jenkins, J. G., Steinbrenner, K., and Bent, D. H., eds., SPSS: Statistical package for the social sciences (second edition): New York, New York, McGraw-Hill, p. 434–467.

Kluth, C. F., 1986, Plate tectonics of the ancestral Rocky Mountains: Tulsa, Oklahoma, American Association of Petroleum Geologists Memoir 41, p. 353–369.

Kolata, D. R., 1975, Middle Ordovician echinoderms from northern Illinois and southern Wisconsin: Lawrence, Kansas, Paleontological Society Memoir 7, 74 p.

Kolata, D. R., 1991, Overview of sequences, in Leighton, M. W., Kolata, D. R., Oltz, D. F., and Eidel, J. J., eds., Interior cratonic basins: Tulsa, Oklahoma, American Association of Petroleum Geologists Memoir 51, p. 59–73.

Kolata, D. R., and Nelson, W. J., 1991, Basin-forming mechanisms of the Illinois Basin, in Leighton, M. W., Kolata, D. R., Oltz, D. F., and Eidel, J. J., eds., Interior cratonic basins: Tulsa, Oklahoma, American Association of Petroleum Geologists Memoir 51, p. 287–292.

Kolata, D. R., and Noger, M. C., 1991, Tippecanoe I subsequence: Middle Ordovician through Lower Devonian Series, in Leighton, M. W., Kolata, D. R., Oltz, D. F., and Eidel, J. J., eds., Interior cratonic basins: Tulsa, Oklahoma, American Association of Petroleum Geologists Memoir 51, p. 89–99.

Kolata, D. R., Frost, J. K., and Huff, W. D., 1986, K-bentonites of the Ordovician Decorah Subgroup, upper Mississippi Valley: Correlation by chemical fingerprinting: Illinois State Geological Survey Circular, 537, 30 p.

Kolata, D. R., Frost, J. K., and Huff, W. D., 1987, Chemical correlation of K-bentonite beds in the Middle Ordovician Decorah Subgroup, upper Mississippi Valley: Geology, v. 15, p. 208–211.

Kolata, D. R., Huff, W. D., and Trevail, R. A., 1990, Correlation of the Ordovician Deicke K-bentonite Bed across the Michigan Basin and southern Ontario: Geological Society of America Abstracts with Programs, v. 22, no. 5, p. 37.

Kreisa, R. D., 1980, The Martinsburg Formation (Middle and Upper Ordovician) and related facies in southwestern Virginia [Ph.D. thesis]: Blacksburg, Virginia Polytechnic Institute and State University, 335 p.

Krekeler, M. P. S., and Huff, W. D., 1993, Occurrence of corrensite and ordered (R3) illite/smectite (I/S) in a VLGM Middle Ordovician K-bentonite from the Hamburg Klippe, central Pennsylvania: Geological Society of America Abstracts with Programs, v. 25, no. 2, p. 30.

Kunk, M. J., and Sutter, J. F., 1984, $^{40}Ar/^{39}Ar$ age spectrum dating of biotite from Middle Ordovician bentonites, eastern North America, in Bruton, D. L., ed., Aspects of the Ordovician System: Palaeontological Contributions of the University of Oslo, no. 295, p. 11–22.

Kunk, M. D., Sutter, J. F., and Bergström, S. M., 1984, $^{40}Ar/^{39}Ar$ age spectrum dating of biotite and sanidine from Middle Ordovician bentonites of Sweden: A comparison with results from eastern North America: Geological Society of America Abstracts with Programs, v. 16, no. 6, p. 566.

Kunk, M. J., Sutter, J. F., Obradovich, J. D., and Lanphere, M. A., 1985, Age of biostratigraphic horizons within the Ordovician and Silurian systems, in Snelling, N. J., ed., The chronology of the geological record: British Geological Survey Memoir 10, p. 89–92.

Kurapkat, W. 1986, Conodont biostratigraphy and paleoecology of the Middle Ordovician ramp-to-basin sequence in the thrust belts of southwest Virginia [M.S. thesis]: Columbus, The Ohio State University, 260 p.

Lagaly, G., 1979, The layer charge of regular interstratified 2:1 clay minerals: Clays and Clay Minerals, v. 27, p. 1–10.

Lagaly, G., and Weiss, A., 1969, Determination of layer charge in mica-type layer silicates: Tokyo, Japan, International Clay Conference 1969, p. 61–80.

Lagaly, G., and Weiss, A., 1976, The layer charge of smectitic layer silicates: Mexico City, Mexico, International Clay Conference 1975, Applied Publishing Ltd., p. 157–172.

Lash, G. G., 1987, Geodynamic evolution of the Lower Paleozoic central Appalachian foreland basin, in Beaumont, C., and Tankard, A. J., eds., Sedimentary basins and basin-forming mechanisms: Calgary, Alberta, Canadian Society of Petroleum Geologists, Memoir 12, p. 412–423.

Laurence, R. A., 1944, An early Ordovician sinkhole deposit of volcanic ash and fossiliferous sediments in east Tennessee: Journal of Geology, v. 52, p. 235–249.

Leighton, M. W., and Kolata, D. R., 1991, Selected interior cratonic basins and their place in the scheme of global tectonics, in Leighton, M. W., Kolata, D. R., Oltz, D. F., Eidel, J. J., eds., Interior cratonic basins: American Association of Petroleum Geologists Memoir 51, p. 729–797.

Leslie, S. A., 1995, Upper Middle Ordovician conodont biofacies and lithofacies distribution patterns in eastern North America and northwestern Europe: Evaluations using the Deicke, Millbrig, and Kinnekulle K-bentonites as time planes [Ph.D. thesis]: Columbus, The Ohio State University, 381 p.

Leslie, S. A., and Bergström, S. M., 1995a, Element morphology and taxonomic relationships of the Ordovician conodonts *Phragmodus primus* Branson and Mehl, 1933, the type species of *Phragmodus* Branson and Mehl, 1933, and *Phragmodus undatus* Branson and Mehl, 1933: Journal of Paleontology, v. 69, no. 5, p. 967–974.

Leslie, S. A., and Bergström, S. M., 1995b, Revision of the North American late Middle Ordovician standard stage classification and timing of the Trenton transgression based on K-bentonite bed correlation, in Cooper, J. D., Droser, M. L., and Finney, S. C., eds., Ordovician odyssey: Fullerton, California, Short papers for the Seventh International Symposium on the Ordovician System, Las Vegas, Nevada, June 1995: Pacific Section SEPM, p. 49–54.

Levorson, C. O., and Gerk, A. J., 1972, A preliminary stratigraphic study of the Galena Group in Winneshiek County, Iowa: Iowa Academy of Science Proceedings, v. 79, p. 111–122.

Levorson, C. O., Gerk, A. J., and Broadhead, T. W., 1979, Stratigraphy of the Dubuque Formation (Upper Ordovician) in Iowa: Iowa Academy of Science Proceedings, v. 86, no. 2, p. 57–65.

Levorson, C. O., Gerk, A. J., Sloan, R. E., and Bisagno, L. A., 1987, General section of the Middle and Late Ordovician strata of northeastern Iowa, in Sloan, R. E., ed., Middle and Late Ordovician lithostratigraphy and biostratigraphy of the upper Mississippi Valley: Minnesota Geological

Survey, Report of Investigations 35, p. 25–39.

Liberty, B. A., 1969, Palaeozoic Geology of the Lake Simcoe Area, Ontario: Geological Survey of Canada, Memoir 355, 201 p.

Liberty, B. A., and Shelden, F. D., 1968, The Geology of Manitoulin Island: Detroit, Michigan Basin Geological Society Annual Field Excursion, 101 p.

Lilienthal, R. T., 1978, Stratigraphic cross-sections of the Michigan Basin: Michigan Department of Natural Resources, Geological Survey Division, Report of Investigation 19, 36 p.

Lipman, P. W., 1966, Water pressures during differentiation and crystallization of some ash-flow magmas from southern Nevada: American Journal of Science, v. 264, p. 810–826.

Loutit, T. S., Hardenbol, J., and Vail, P. R., 1988, Condensed sections: the key to age determination and correlation of continental margin sequences, in Wilgus, C. K., Posamentier, H., Ross, C. A., Kendall, C. G. St. C., eds., Sea-level changes: An integrated approach: Tulsa, Oklahoma, Society of Economic Paleontologists and Mineralogists, Special Publication 42, p. 183–213.

MacKenzie, P., and Bergström, S. M., 1993, Discovery of the zonal index conodont *Amorphognathus ordovicicus* in the Richmondian of Indiana: Implications for the regional correlation of the North American Upper Ordovician standard: Geological Society of America Abstracts with Programs, v. 25, no. 6, p. A–472.

Malpas, J., and Stevens, R. K., 1977, The origin and emplacement of the ophiolite suite with examples from western Newfoundland: Geotectonics, v. 11, p. 453–466.

McCracken, E., 1955, Correlation of insoluble residue of upper Arbuckle of Missouri and southern Kansas: American Association of Petroleum Geologists Bulletin, v. 39, no. 1, p. 47–59.

McFarlan, A. C., 1943, Geology of Kentucky: Lexington, Kentucky, University of Kentucky, 531 p.

McKerrow, W. S., Dewey, J. F., and Scotese, C. R., 1991, The Ordovician and Silurian development of the Iapetus Ocean: The Palaeontological Association Special Papers in Palaeontology, no. 44, p. 165–178.

McVey, D. E., 1993, Timing of the Blount and Martinsburg foreland basin development during the Taconic orogeny based on the Deicke and Millbrig K-bentonite marker horizons [M.S. thesis]: Cincinnati, Ohio, University of Cincinnati, 133 p.

Milici, R. C., 1969, Middle Ordovician stratigraphy in central Sequatchie Valley, Tennessee: Southeastern Geology, v. 11, p. 111–128.

Milici, R. C., and Smith, J. W., 1969, Stratigraphy of the Chickamauga Supergroup in its type area: Georgia Geological Survey, Bulletin 80, p. 1–35.

Miller, R. L., 1937, Stratigraphy of the Jacksonburg Limestone: Geological Society of America Bulletin, v. 48, no. 11, p. 1687–1718.

Miller, R. L., and Brosgé, W. P., 1954, Geology and oil resources of the Jonesville district, Lee County, Virginia: U.S. Geological Survey Bulletin 990, 240 p.

Miller, R. L., and Fuller, J. O., 1954, Geology and oil resources of the Rose Hill district—The fenster area of the Cumberland overthrust block—Lee County, Virginia: Virginia Geological Survey Bulletin 71, 383 p.

Mitchell, C. E., Goldman, D., Delano, J. W., Samson, S. D., and Bergström, S. M., 1994, Temporal and spacial distribution of biozones and facies relative to geochemically correlated K-bentonites in the Middle Ordovician Taconic foredeep: Geology, v. 22, p. 715–718.

Mossler, J. H., 1985, Sedimentology of the Middle Ordovician Platteville Formation southeastern Minnesota: Minnesota Geological Survey, Report of Investigations 33, 27 p.

Mossler, J. H., and Hayes, J. B., 1966, Ordovician potassium bentonites of Iowa: Journal of Sedimentary Petrology, v. 36, no. 2, p. 414–427.

Nadeau, P. H., 1985, The physical dimensions of fundamental clay particles: Clay Minerals, v. 20, p. 499–514.

Nadeau, P. H., and Bain, D. C., 1986, Composition of some smectites and diagenetic illitic clays and implications for their origin: Clays and Clay Minerals, v. 34, p. 455–464.

Nadeau, P. H., Tait, J. M., McHardy, W. J., and Wilson, M. J., 1984, Interstratified XRD characteristics of physical mixtures of elementary clay particles: Clay Minerals, v. 19, p. 67–76.

Nadeau, P. H., Wilson, M. J., McHardy, W. J., and Tait, J. M., 1985, The conversion of smectite to illite during diagenesis: Evidence from some illitic clays from bentonites and sandstones: Mineralogical Magazine, v. 49, p. 393–400.

Nelson, W. A., 1921, Notes on a volcanic ash bed in the Ordovician of middle Tennessee: Tennessee Geological Survey Bulletin, v. 25, p. 46–48.

Nelson, W. A., 1922, Volcanic ash beds in the Ordovician of Tennessee, Kentucky and Alabama: Geological Society of America Bulletin, v. 33, p. 605–615.

Nelson, W. A., 1925, Two New Volcanic Ash Horizons in the Stones River Group of the Ordovician of Tennessee, Pan-American Geologist, v. 43, no. 2, p. 513.

Nelson, W. A., 1926, Volcanic ash deposit in the Ordovician of Virginia: Pan-American Geologist, v. 45, no. 1, p. 96.

North American Commission on Stratigraphic Nomenclature, 1983, North American Stratigraphic Code: American Association of Petroleum Geologists Bulletin, v. 67, p. 841–875.

O'Neill, B. E., 1985, Conodont biostratigraphy and paleoecology in the lower Middle Ordovician of the Valley and Ridge overthrust region of west-central Virginia [M.S. thesis]: Columbus, The Ohio State University, 194 p.

Patchen, D. G., Avary, K. L., and Erwin, R. B., 1985, Northern Appalachian region, in Lindberg, F. A., ed., Correlation of stratigraphic units of North America: American Association of Petroleum Geologists, COSUNA Chart Series MBA.

Pearce, J. A., 1982, Trace element characteristics of lavas from destructive plate boundaries, in Thorpe, R. S., ed., Andesites: New York, Wiley, p. 525–548.

Pearce, J. A., Harris, N. B. W., and Tindle, A. G., 1984, Trace element discrimination diagrams for the tectonic interpretation of granitic rocks: Journal of Petrology, v. 25, p. 956–983.

Perry, W. J., Jr., 1964, Geology of the Ray Sponaugle well: American Association of Petroleum Geologists Bulletin, v. 48, p. 659–669.

Perry, W. J., Jr., 1972, The Trenton Group of Nittany Anticlinorium, eastern West Virginia: West Virginia Economic and Geological Survey Circular Series no. 13, 30 p.

Prouty, C. E., 1959, The Annville, Myerstown and Hershey Formations of Pennsylvania: Pennsylvania Geological Survey, 4th Series, Bulletin G31, 47 p.

Purser, B. H., 1969, Syn-sedimentary marine lithification of Middle Jurassic limestones in the Paris Basin: Sedimentology, v. 12, p. 205–230.

Quinlan, G. M., and Beaumont, C., 1984, Appalachian thrusting, lithospheric flexure, and the Paleozoic stratigraphy of the eastern interior of North America: Canadian Journal of Earth Sciences, v. 21, p. 973–996.

Rader, E. K., and Read, J. F., 1989, Early Paleozoic continental shelf to basin transition, northern Virginia, in International Geological Congress, 28th, Field Trip Guidebook T221: Washington, D.C., American Geophysical Union, p. 1–9.

Rampino, M. R., and Self, S., 1993, Climate-volcanism feedback and the Toba eruption of ~74,000 years ago: Quaternary Research, v. 40, p. 269–280.

Repetski, J. E., 1973, The conodont fauna of the Dutchtown Formation (Middle Ordovician) of southeast Missouri [M.S. thesis]: Columbia, University of Missouri, 182 p.

Reynolds, R. C., 1980, Interstratified clay minerals, in Brindley, G. W., and Brown, G., eds., Crystal structures of clay minerals and their X-ray identification: London, United Kingdom, Mineralogical Society, p. 249–303.

Reynolds, R. C., Jr., 1985, NEWMOD: A computer program for the calculation of one-dimensional diffraction patterns of mixed-layer clays: Hanover, New Hampshire, R. C. Reynolds (8 Brook Road).

Reynolds, R. C., Jr., 1990, A preliminary study of order/disorder and polytypism in mixed-layered illite/smectite: Clay Minerals Society, 27th Annual Meeting, Columbia, Missouri, Program and Abstracts, p. 105.

Reynolds, R. C., Jr., 1993, Three-dimensional X-ray powder diffraction from disordered illite: Simulation and interpretation of the diffraction patterns, *in* Reynolds, R. C., Jr., and Walker, J. R., eds., Computer applications to X-ray powder diffraction analysis of clay minerals: Workshop lectures, v. 5: Boulder, Colorado, Clay Minerals Society, p. 44–78.

Reynolds, R. C., Jr., and Hower, J., 1970, The nature of interlayering in mixed-layer illite-montmorillonites: Clays and Clay Minerals, v. 18, p. 25–36.

Riva, J., 1972, Geology of the environs of Quebec City, International Geological Congress, 24th, Guidebook for Excursion B-19: International Geophysical Union, 53 p.

Roberts, B., and Merriman, R. J., 1990, Cambrian and Ordovician metabentonites and their relevance to the origins of associated mudrocks in the northern sector of the Lower Palaeozoic Welsh marginal basin: Geological Magazine, v. 127, p. 31–43.

Rodgers, J., 1953, Geologic map of east Tennessee with explanatory note: Tennessee Division of Geology Bulletin 58, Part II, 168 p.

Rodgers, J., 1987, The Appalachian-Ouachita orogenic belt: Episodes, v. 10, p. 259–266.

Rones, M., 1969, A lithostratigraphic, petrographic, and chemical investigation of the lower Middle Ordovician carbonate rocks in central Pennsylvania: Pennsylvania Geological Survey, General Geology Report G53, 224 p.

Rosenkrans, R. R., 1933, Bentonite in northern Virginia: Washington Academy of Science Journal, v. 23, no. 9, p. 413–419.

Rosenkrans, R. R., 1934a, Correlation studies of the central and south-central Pennsylvania bentonite occurrences: American Journal of Science, v. 17, p. 113–134.

Rosenkrans, R. R., 1934b, Some problems involved in bentonite studies, *in* Proceedings, Annual Meeting, Geological Society of America, Chicago, 1933: New York, New York, Geological Society of America, p. 380.

Rosenkrans, R. R., 1936, Stratigraphy of Ordovician bentonite beds in southwestern Virginia: Virginia Geological Survey Bulletin, v. 46-I, p. 85–111.

Ross, C. S., 1928, Altered Paleozoic volcanic materials and their recognition: American Association of Petroleum Geologists Bulletin, v. 12, p. 143–164.

Ross, C. S., and Kerr, P. F., 1934, Bentonite and related clays: Geological Society of America Proceedings 1933, p. 380.

Ross, C. S., and Shannon, E. V., 1925, Nature of bentonite and related clays: Pan-American Geologist, v. 43, no. 5, p. 364–365.

Ross, C. S., and Shannon, E. V., 1926, The minerals of bentonite and related clays and their physical properties: Journal of American Ceramic Society, v. 9, no. 2, p. 77–96.

Ross, R. J., Jr., 1976, Ordovician sedimentation in the western United States, *in* Bassett, M. G., ed., The Ordovician System, Proceedings of the Palaeontological Association Symposium, Birmingham, England, 1974, University of Wales Press, Cardiff, p. 73–105.

Ross, R. J., Jr., Naeser, C. W., Bergström, S. M., and Cressman, E. R., 1981, Bentonites in the Tyrone Limestone dated by fission tracks: Geological Society of America Abstracts with Programs, v. 13, p. 541.

Ross, R. J., Jr., and 14 others, 1982, Fission track dating of British Ordovician and Silurian stratotypes: Geological Magazine, v. 119, p. 135–153.

Rowley, D. B., and Kidd, W. S. F., 1981, Stratigraphic relationships and detrital composition of the medial Ordovician flysch of western New England: Implications for the tectonic evolution of the Taconic Orogeny: Journal of Geology, v. 89, p. 199–218.

Ruedemann, R., 1912, The Lower Siluric shales of the Mohawk Valley: New York State Museum, Bulletin 162, 151 p.

Ruppel, S. C., and Walker, K. R., 1977, The ecostratigraphy of the Middle Ordovician of the southern Appalachians (Kentucky, Tennessee, and Virginia), U.S.A.: A field excursion: University of Tennessee Department of Geological Sciences, Studies in Geology, no. 77-1.

Ryder, R. T., 1991, Stratigraphic framework of Cambrian and Ordovician rocks in the central Appalachian basin from Richland County, Ohio, to Rockingham County, Virginia: U.S. Geological Survey Miscellaneous Investigations Series Map I-2264.

Ryder, R. T., 1992a, Stratigraphic framework of Cambrian and Ordovician rocks in the central Appalachian basin from Lake County, Ohio, to Juniata County, Pennsylvania: U.S. Geological Survey Miscellaneous Investigations Series Map I-2200.

Ryder, R. T., 1992b, Stratigraphic framework of Cambrian and Ordovician rocks in the central Appalachian basin from Morrow County, Ohio, to Pendelton County, West Virginia: U.S. Geological Survey Bulletin 1839-G, 25 p.

Ryder, R. T., Harris, A. G., Repetski, J. E., 1992, Stratigraphic framework of Cambrian and Ordovician rocks in the central Appalachian basin from Medina County, Ohio, through southwestern and south-central Pennsylvania to Hampshire County, West Virginia: U.S. Geological Survey Bulletin 1839-K, 32 p.

Samson, S. D., Kyle, P. R., and Alexander, E. C., Jr., 1988, Correlation of North American Ordovician bentonites by using apatite chemistry: Geology, v. 16, no. 5, p. 444–447.

Samson, S. D., Patchett, P. J., Roddick, J. C., and Parrish, R. R., 1989, Origin and tectonic setting of Ordovician bentonites in North America: Isotopic and age constraints: Geological Society of America Bulletin, v. 101, p. 1175–1181.

Sanford, B. V., 1961, Subsurface stratigraphy of Ordovician rocks in southwestern Ontario: Geological Survey of Canada, Paper 60-26, 54 p.

Sardeson, F. W., 1924, Volcanic ash in Ordovicic rocks of Minnesota: Pan-American Geologist, v. 42, p. 45–52.

Sardeson, F. W., 1926a, Beloit Formation and bentonite: Pan-American Geologist, v. 45, p. 383–392 and v. 46, p. 11–24.

Sardeson, F. W., 1926b, Pioneer repopulation of devastated sea-bottoms: Pan-American Geologist, v. 46, p. 273–288.

Sardeson, F. W., 1927, Ordovicic bentonite in the northwest: Pan-American Geologist, v. 48, p. 347–354.

Sardeson, F. W., 1928, Bentonite seams in stratigraphic correlation: Pan-American Geologist, v. 50, p. 107–116.

Sardeson, F. W., 1934, Ordovicic bentonite zones: Pan-American Geologist, v. 61, p. 19–28.

Schmidt, M. A., 1982, Conodont biostratigraphy and facies relations of the Chickamauga Limestone (Middle Ordovician) of the Southern Appalachians, Alabama and Georgia, [M.S. thesis]: Columbus, The Ohio State University.

Schumacher, G. A., and Carlton, R. W., 1991, Impure K-bentonite beds from the Lexington Limestone and the Point Pleasant Formation (Middle Ordovician) of northern Kentucky and southwestern Ohio: Southeastern Geology, v. 32, no. 2, p. 83–105.

Secor, D. T., Samson, S. L., Snoke, A. W., and Palmer, A. R., 1983, Confirmation of the Carolina slate belt as an exotic terrane: Science, v. 221, p. 649–651.

Shanmugam, G., and Lash, G. G., 1982, Analogous tectonic evolution of the Ordovician foredeeps, southern and central Appalachians: Geology, v. 10, p. 562–566.

Sherwood, W. C., 1964, Structure of the Jacksonburg Formation in Northampton and Lehigh Counties, Pennsylvania: Pennsylvania Geological Survey, General Geology Report, G 45, 64 p.

Sloan, R. E., 1987, Tectonics, biostratigraphy, and lithostratigraphy of the Middle and Late Ordovician of the upper Mississippi Valley, *in* Sloan, R. E., ed., Middle and Late Ordovician lithostratigraphy and biostratigraphy of the upper Mississippi Valley: Minnesota Geological Survey Report of Investigations 35, p. 7–20.

Smith, R. C., Way, J. H., and Berkheiser, S. W., Jr. 1986, Road log day 1, 51st Annual Field Conference of Pennsylvania Geologists: Selected geology of Bedford and Huntingdon Counties, Guidebook: Harrisburg, Pennsylvania, p. 110–119.

Spencer, A. C., Kummel, H. B., Wolff, J. E., Salisburg, R. D., and Palache, C., 1908, Description of Franklin Furnace quadrangle, New Jersey: U.S. Geological Survey Atlas, no. 161, Franklin Furnace folio, 27 p.

Środoń, J., 1980, Precise identification of illite/smectite interstratifications by X-ray powder diffraction: Clays and Clay Minerals, v. 28, p. 401–411.

Środoń, J., Morgan, D. J., Eslinger, E. V., Eberl, D. D., and Karlinger, M. R., 1986, Chemistry of illite/smectite and end-member illite: Clays and Clay Minerals, v. 34, p. 368–378.

Środoń, J., Elsass, F., McHardy, W. J., and Morgan, D. J., 1992, Chemistry of illite-smectite inferred from TEM measurements of fundamental particles: Clay Minerals, v. 27, p. 137–158.

Stith, D. A., 1979, Chemical composition, stratigraphy, and depositional environments of the Black River Group (Middle Ordovician), southwestern Ohio: Ohio Division of Geological Survey, Report of Investigations no. 113, 36 p.

Stith, D. A., 1986, Supplemental core investigations for high-calcium limestones in western Ohio and discussion of natural gas and stratigraphic relationships in the Middle to Upper Ordovician rocks of southwestern Ohio: Ohio Division of Geological Survey, Report of Investigations no. 132, 17 p.

Stose, G. W., and Jonas, A. I., 1927, Ordovician shale and associated lava in southeastern Pennsylvania: Geological Society of America Bulletin, v. 38, p. 505–536.

Sweet, W. C., 1984, Graphic correlation of upper Middle and Upper Ordovician rocks, North American Midcontinent Province, U.S.A., in Bruton, D. L., ed., Aspects of the Ordovician System: Palaeontological Contributions from the University of Oslo, no. 295, p. 23–35.

Sweet, W. C., 1987, Distribution and significance of conodonts in Middle and Upper Ordovician strata of the Upper Mississippi Valley region, in Sloan, R. E., ed., Middle and Late Ordovician lithostratigraphy and biostratigraphy of the Upper Mississippi Valley: Minnesota Geological Survey Report of Investigations 35, p. 167–172.

Sweet, W. C., 1988, Mohawkian and Cincinnatian chronostratigraphy: New York State Museum Bulletin N 462, p. 84–90.

Sweet, W. C., 1995, A conodont-based composite standard for the North American Ordovician: Progress Report, in Cooper, J. D., Droser, M. L., and Finney, S. C., eds., Ordovician odyssey: Short papers for the Seventh International Symposium on the Ordovician System, Las Vegas, Nevada, June 1995: Fullerton, California, SEPM, p. 15–20.

Sweet, W. C., and Bergström, S. M., 1976, Conodont biostratigraphy of the Middle and upper Ordovician of the United States Midcontinent, in Bassett, M. G., ed., The Ordovician system, in Proceedings, Palaeontological Association Symposium, Birmingham, U.K., September 1974: Cardiff, University of Wales Press, p. 121–151.

Sweet, W. C., and Bergström, S. M., 1986, Conodonts and biostratigraphic correlation: Annual Review of Earth and Planetary Sciences, v. 14, p. 85–112.

Sweet, W. C., Ethington, R. L., and Barnes, C. R., 1971, North American Middle and Upper Ordovician conodont faunas: Boulder, Colorado, Geological Society of America Memoir 127, p. 163–193.

Taylor, H., 1947, Middle Ordovician limestones in central Kansas: American Association of Petroleum Geologists Bulletin, v. 31, no. 7, p. 1242–1282.

Templeton, J. S., and Willman, H. B., 1963, Champlainian series (Middle Ordovician) in Illinois: Illinois State Geological Survey, Bulletin 89, 260 p.

Thomas, W. A., 1972, Regional Paleozoic stratigraphy in Mississippi between Ouachita and Appalachian Mountains: American Association of Petroleum Geologists Bulletin, v. 56, p. 81–106.

Thomas, W. A., 1988, The Black Warrior Basin, in Sloss, L. L., ed., Sedimentary cover—North American Craton: U.S.: Boulder, Colorado, Geological Society of America, Geology of North America, v. D-2, p. 471–492.

Thompson, R. R., 1963, Lithostratigraphy of the Middle Ordovician Salona and Coburn Formations in central Pennsylvania: Pennsylvania Geological Survey, General Geology Report 38, 154 p.

Thompson, T. L., 1991, Paleozoic Succession in Missouri, Part 2, Ordovician System: Missouri Geological Survey Report of Investigations 70, 282 p.

Trevail, R. A., 1988, Distribution of Middle Ordovician altered volcanic ash beds (K-bentonites), southwestern Ontario: Geological Association of Canada/Mineralogical Association of Canada/Canadian Society of Petroleum Geologists, Program with Abstracts, v. 13, p. A126.

Trevail, R. A., 1990, Ordovician K-bentonites in the subsurface of southwestern Ontario, in Proceedings, Ontario Petroleum Institute, 29th Annual Conference: Ontario Petroleum Institute, 18 p.

Trevail, R. A., Huff, W. D., and Kolata, D. R., 1989, Facies interpretation of Ordovician carbonates of southern Ontario using K-bentonites [abs.]: American Association of Petroleum Geologists Bulletin, v. 73, p. 1040.

Treworgy, J. D., Whitaker, S. T., and Lasemi, Z., 1994, 11:30 o'clock cross section in the Illinois Basin, Wayne County to Stephenson County, Illinois: Illinois State Geological Survey Open File Series 1994-6, 13 p.

Tucker, R. D., Krogh, T. E., Ross, R. J., Jr., and Williams, S. H., 1990, Timescale calibration by high-precision U-Pb zircon dating of interstratified volcanic ashes in the Ordovician and Lower Silurian stratotypes of Britain: Earth and Planetary Science Letters, v. 100, p. 51–58.

Ulrich, E. O., 1888, Correlation of the Lower Silurian horizons of Tennessee and of the Ohio and Mississippi valleys with those of New York and Canada: American Geologist, v. 1, p. 100–110.

Ulrich, E. O., and Everett, O., 1890, Descriptions of Lower Silurian sponges, in Worthen, A. W., ed., Geology and paleontology of Illinois, v. 8, p. 209–241.

Vasseur, G., and Velde, B., 1993, A kinetic interpretation of the smectite-to-illite transformation, in Dore, A.G., ed., Basin modelling: Advances and applications: Amsterdam, The Netherlands, Elsevier, Norwegian Petroleum Society (NPF) Special Publication 3, p. 173–184.

Veblen, D. R., Guthrie, G. D., Livi, K. J. T., and Reynolds, R. C., 1990, High-resolution transmission electron microscopy and electron diffraction of mixed-layer illite/smectite: Experimental results: Clays and Clay Minerals, v. 38, p. 1–13.

Velde, B., and Brusewitz, A. M., 1982, Metasomatic and non-metasomatic low grade metamorphism of Ordovician meta-bentonites in Sweden: Geochimica et Cosmochimica Acta, v. 46, p. 447–452.

Votaw, R. B., 1971, Conodont biostratigraphy of the Black River Group (Middle Ordovician) and equivalent rocks of the eastern Midcontinent, North America [Ph.D. thesis]: Columbus, The Ohio State University.

Votaw, R. B., 1980, Ordovician and Silurian geology of the northern peninsula of Michigan: Detroit, Michigan Basin Geological Society Field Conference.

Wagner, W. R., 1961, Subsurface Cambro-Ordovician stratigraphy of northwestern Pennsylvania and bordering states: Pennsylvania Geological Survey, Fourth Series, Progress Report 156, 22 p.

Walker, K. R., 1973, Stratigraphy and environmental sedimentology of Middle Ordovician Black River Group in the type area—New York State: New York State Museum and Science Service, Bulletin 419, 43 p.

Wang, P., and Glover, L., 1992, A tectonics test of the most commonly used geochemical discriminant diagrams and patterns: Earth-Science Reviews, v. 33, p. 111–131.

Watanabe, T., 1988, The structural model of illite/smectite interstratified minerals and the diagram for its identification: Clay Science, v. 7, p. 97–114.

Weaver, C. E., 1953, Mineralogy and petrology of some Ordovician K-bentonites and related limestones: Geological Society of America Bulletin, v. 64, p. 921–943.

Weaver, C. E., 1956, The distribution and identification of mixed-layer clays in sedimentary rocks: American Mineralogist, v. 41, p. 202–221.

Weiss, M. P., 1954, Feldspathized shales from Minnesota: Journal of Sedimentary Petrology, v. 24, no. 4, p. 270–274.

Weiss, M. P., 1955, Some Ordovician brachiopods from Minnesota and their stratigraphic relations: Journal of Paleontology, v. 29, p. 759–774.

Weiss, M. P., 1957, Upper Middle Ordovician stratigraphy of Fillmore County, Minnesota: Geological Society of America Bulletin, v. 68, p. 1027–1062.

Weiss, M. P., and Bell, W. C., 1956, Middle Ordovician rocks of Minnesota and their lateral relations: Geological Society of America Guidebook Field Trip No. 2, Minneapolis, Minnesota Meeting, p. 55–73.

Westrich, H. R., and Gerlach, T. M., 1992, Magmatic gas source for the stratospheric SO_2 cloud from the June 15, 1991, eruption of Mount Pinatubo: Geology, v. 20, p. 867–870.

Whitcomb, L., 1932, Correlation by Ordovician bentonite: Journal of Geology, v. 40, p. 522–534.

Whitcomb, L., 1934, Possible volcanic sources of the Ordovicic bentonites, in Proceedings, Geological Society of America, 1933: New York, New York, Geological Society of America, p. 382.

Whitcomb, L., 1935, Possible volcanic sources of Ordovicic bentonites: Pan-

American Geologist, v. 63, no. 4, p. 265–270.

Whitcomb, L., and Rosenkrans, R. R., 1934, Bentonites in the Lower Chambersburg, *in* Proceedings, Annual Meeting, Geological Society of America, Chicago, 1933: New York, New York, Geological Society of America, p. 381.

White, W. H., 1989, Geochemical evidence for crust-to-mantle recycling in subduction zones, *in* Hart, S. R., and Gulen, L., eds., Crust/mantle recycling at convergence zones: Amsterdam, The Netherlands, Kluwer, p. 43–58.

Wickstrom, L. H., Botoman, G., and Stith, D. A., 1985, Report on a continuously cored hole drilled into the Precambrian in Seneca County, northwestern Ohio: Ohio Division of Geological Survey, Information Circular, no. 51.

Wickstrom, L. H., Gray, J. D., Stieglitz, R. D., 1992, Stratigraphy, structure, and production history of the Trenton Limestone (Ordovician) and adjacent strata in northwestern Ohio: Ohio Division of Geological Survey, Report of Investigations, no. 143, 78 p.

Willman, H. B., and Buschbach, T. C., 1975, Ordovician System, *in* Willman, H. B., Atherton, A., Buschbach, T. C., Collinson, C., Frye, J. C., Hopkins, M. E., Lineback, J. A., and Simon, J. A., eds., Handbook of Illinois Stratigraphy: Illinois State Geological Survey Bulletin 95, p. 47–87.

Willman, H. B., and Kolata, D. R., 1978, The Platteville and Galena Groups in northern Illinois: Illinois State Geological Survey Circular 502, 75 p.

Wilson, C. W., Jr., 1949, Pre-Chattanooga stratigraphy in central Tennessee: Tennessee Department of Conservation Division of Geology Bulletin 56, 407 p.

Wilson, C. W., Jr., 1962, Stratigraphy and geologic history of Middle Ordovician rocks of central Tennessee: Geological Society of America Bulletin, v. 73, p. 481–504.

Winchester, J. A., and Floyd, P. A., 1977, Geochemical discrimination of different magma series and their differentiation products using immobile elements: Chemical Geology, v. 20, p. 325–343.

Witzke, B. J., 1980, Middle and Upper Ordovician paleogeography of the region bordering the Transcontinental Arch, *in* Fouch, T. D., and Magathan, E. R., eds., Paleozoic paleogeography of west-central United States—west-central United States Paleogeography Symposium 1: Society of Economic Paleontologists and Mineralogists, p. 1–18.

Witzke, B. J., and Kolata, D. R., 1988, Changing structural and depositional patterns, Ordovician Champlainian and Cincinnatian Series of Iowa-Illinois, *in* Ludvigson, G. A., and Bunker, B. J., eds., New perspectives on the Paleozoic history of the Upper Mississippi Valley—An examination of the Plum River Fault Zone: Iowa Department of Natural Resources Geological Survey Bureau, Guidebook no. 8, p. 55–71.

Wood, D. A., Joron, J. L., and Treuil, M., 1979, A re-appraisal of the use of trace elements to classify and discriminate between magma series erupted in different tectonic settings: Earth and Planetary Science Letters, v. 45, p. 326–336.

Woodard, H. H., 1972, Syngenetic sanidine beds from Middle Ordovician St. Peter Sandstone, Wisconsin: Journal of Geology, v. 80, p. 323–332.

Yost, D. A., Huff, W. D., Bergström, S. M., and Kolata, D. R., 1994, Use of mineralogical and geochemical data for high resolution stratigraphic correlation of a Middle Ordovician K-bentonite: Geological Society of America Abstracts with Programs, v. 26, no. 3, p. 81.

Young, D. M., 1940, Bentonitic clay horizons and associated chert layers of central Kentucky: University of Kentucky Research Club Bulletin 6, p. 27–31.

Young, L. M., 1970, Early Ordovician sedimentary history of Marathon geosyncline, Trans-Pecos, Texas: American Association of Petroleum Geologists Bulletin, v. 54, p. 2303–2316.

Zhang, Y. S., and Huff, W. D., 1994, Grain size characteristics of a Middle Ordovician eruption and its source area, *in* Abstracts of the Eighth International Conference on Geochronology, Cosmochronology and Isotope Geology, U.S. Geological Survey Circular 1107, p. 364.

Ziegler, P. A., 1988, Evolution of the Arctic–North Atlantic and the Western Tethys: Tulsa, Oklahoma, American Association of Petroleum Geologists Memoir 43, 198 p.

MANUSCRIPT ACCEPTED BY THE SOCIETY JANUARY 24, 1996